FOUNDATIONS OF ENGINEERING

FOUNDATIONS OF ENGINEERING

Developing the Skills and Mindset of a Successful Engineer

FIRST EDITION

Tanya Kunberger
Florida Gulf Coast University

R. Christopher Geiger
Florida Gulf Coast University

Bassim Hamadeh, CEO and Publisher
Jennifer McCarthy, Senior Field Acquisitions Editor
Anne Jones, Project Editor
Alia Bales, Associate Production Manager
Emely Villavicencio, Senior Graphic Designer
Trey Soto, Licensing Specialist
Natalie Piccotti, Director of Marketing
Kassie Graves, Senior Vice President, Editorial
Jamie Giganti, Director of Academic Publishing

Copyright © 2023 by Cognella, Inc. All rights reserved. No part of this publication may be reprinted, reproduced, transmitted, or utilized in any form or by any electronic, mechanical, or other means, now known or hereafter invented, including photocopying, microfilming, and recording, or in any information retrieval system without the written permission of Cognella, Inc. For inquiries regarding permissions, translations, foreign rights, audio rights, and any other forms of reproduction, please contact the Cognella Licensing Department at rights@cognella.com.

Trademark Notice: Product or corporate names may be trademarks or registered trademarks and are used only for identification and explanation without intent to infringe.

Cover image cover image copyright © 2016 iStockphoto LP/Ed-blitzkrieg.

Icon images:
Source: https://openclipart.org/detail/32353.
Source: http://www.clker.com/clipart-2312.html.
Source: http://www.clker.com/clipart-2985.html.
Source: http://www.clker.com/clipart-gears-1.html.

Printed in the United States of America.

CONTENTS

Introduction   viii

CHAPTER 1   DESIGN PROCESS   XIV

Before You Begin   xiv
Learning Objectives   xiv
Introduction   xiv
Problem Identification—Observations   3
Defining Criteria and Constraints   7
Gathering Data   9
Making and Justifying Assumptions   10
Brainstorming   12
Analyzing Potential Solutions   14
Testing and Validating Your Solution   17
Communicate Your Findings   19
Innovation   20
Entrepreneurial Mindset   22
Chapter Summary   23
Chapter References   25

CHAPTER 2   ENGINEERING SUCCESS   26

Before You Begin   26
Learning Objectives   26
Introduction   26
Metacognition and Growth Mindset   27
Plan for Success/Setting Goals   28
Organization and Time Management   32
Classes   35
Academic Integrity   37
Leadership/Teamwork   39

Campus Engagement   42
Faculty/Advisor Relationships   45
Networking   47
Chapter Summary   49
Chapter References   51

CHAPTER 3   ENGINEERING CAREERS   52

Before You Begin   52
Learning Objectives   52
Introduction   52
Résumés   55
Cover Letters   63
Electronic Résumés   64
Portfolios and Electronic Portfolios   65
Interviews   66
Internships, Co-ops, and REUs   69
Graduate School   73
Campus Resources   75
Chapter Summary   76
Chapter References   78

CHAPTER 4   INFORMATION LITERACY   80

Before You Begin   80
Learning Objectives   80
Introduction   80
Current Resources   84
Reading Scholarly Articles   94
Synthesizing Information   97
Plagiarism   98
Reference Managers and Citations   101
Chapter Summary   105
Chapter References   106

CHAPTER 5   WRITTEN COMMUNICATION   108

Before You Begin   108
Learning Objectives   108
Introduction   108

Writing Clarity   109
Types of Written Communication   117
Defining Variables and Abbreviations   124
Including Figures, Tables, and Equations   125
Significant Figures   134
Statistics   136
Appendices   139
Chapter Summary   140
Chapter References   143

CHAPTER 6   ORAL COMMUNICATION   144

Before You Begin   144
Learning Objectives   144
Introduction   144
Speed, Tone, and Volume   146
Level of Formality   151
Video Conferencing   153
Posters and Slide Presentations   156
Chapter Summary   161
Chapter References   163

CONCLUSION   164

The End (and Beginning) of Your Journey   164
Engineering Success Starts with You   167
References   170

INTRODUCTION

BEFORE YOU BEGIN

Take a minute to consider the following questions:

- What information do you expect to obtain from an introductory engineering textbook and/or an introductory engineering course?

- How do you define the term "ethics"?

- When and how should an engineer be ethical?

LEARNING OBJECTIVES

By the end of the introduction, you should be able to:

- Summarize the layout of this book and what topics are covered.

- Define how engineering ethics plays a role in the actions of both engineering students and practicing engineers.

OVERVIEW

Welcome to the preliminary edition of *Foundations of Engineering: Developing the skills and mindset of a successful engineer*. When we set out to write this textbook, we wanted to have something that worked well with our one-credit Introduction to the Engineering Profession course at our home institution. As this is a course taken by all of our engineering disciplines, we have tried to avoid discipline-specific content to make this text as accessible and relevant to as many disciplines as possible (although we fully admit it is possible our civil engineering and bioengineering backgrounds may result in some of our examples slanting toward those disciplines). Additionally, we were unable to find a single textbook that encompassed the breadth of content we focused on in our course: the broad basis of the engineering design process, information about student success and career development, and the communication skills necessary to be successful while in school and beyond. For students or instructors looking for discipline specific information or perhaps a text that focuses on fundamental skills associated with engineering, this book may not be sufficient to aid in the learning within those concentration areas.

As suggested in the previous paragraph, we cover three main topic areas over the course of this book. Chapter 1 discusses in very general terms the engineering design process; in addition to focusing on the steps of that process we also touch upon the concepts of innovation and entrepreneurship (although in the form of mindset vs. what it is to be an entrepreneur). Chapters 2 and 3 focus on setting up a path for success as a student, and how that should translate to future success in your chosen career path. We consider activities, methods and relationships that will not only help you in the present, but if carried through can help you achieve future goals and aspirations. Additionally, within these chapters we consider future career trajectories and the tools necessary to effectively begin searching for positions that align you with those trajectories. Finally, in Chapters 4–6, we focus on communication, first with respect to how to find the information you need (Chapter 4), then how to disseminate that information in both written (Chapter 5) and oral (Chapter 6) formats.

Throughout this book, you will find that the general format seen within this chapter is carried across each of the remaining chapters. It is our anticipation that this textbook will not necessarily be read cover to cover, but specific chapters or even specific sections will be used in a course or will be valued by a student. Each chapter can in many ways stand on its own, and any recurrence of content outside of the chapter in question will be referenced or referred to appropriately. Even within a given chapter, the intent of the textbook is not to have the reader passively read the text from beginning to end. We have purposefully included questions for consideration prior to the start of each chapter and have also laid out a series of learning objectives that readers should feel have been adequately addressed by the end of the chapter. In addition, within the chapter, activity call-outs have been included to reinforce key concepts that have been discussed. These call-outs can be used as in-class activities for courses that have adopted this book as part of their materials, or can simply be completed by the reader on her/his/their own as a means of individual reinforcement. Additionally, each chapter includes a set of end-of-chapter activities that can be used in a similar manner. These activities are typically more involved than the in-text call-outs, and may be appropriate for longer, more robust self-reflection by the reader and potentially included as part of the assessment activities for a course.

One last thing we wanted to address in this introduction is the recognition that the topics covered in this book can have multiple interpretations, and that our discussions and viewpoints are simply one such interpretation. Unlike math or other aspects of engineering, where 2+2 = 4 or $\Sigma Fx = 0$ for an object in static equilibrium, topics such as the steps of the engineering design process, resumes, and technical communication have more nuanced answers that provide an opportunity for disagreement or alternative interpretations. The methods we describe in this text come from a combined 25+ years of teaching both introductory and upper level design-based courses. Because there is not one right interpretation of these topics, our own interpretation over the years has changed and become more refined as well. Certainly some of that refinement is a function of simply how times have changed (for example, it did take us a while to convert from two spaces after a period to just one[1]), but some of it is also a result of changes in our own ways of thinking (hopefully for the better). Regardless, please recognize that if what is written in this book contradicts your own way of thinking or information presented within a class, you can take away from that conflict that both points of view might very well be correct, and it is up to you to determine how (or even if) those different viewpoints can be of value as you develop your own sense of understanding of the topic.

Many of the topics covered in this book can have multiple interpretations that open up the opportunity for discussion and the development of your own sense of understanding of the topic.

ENGINEERING ETHICS

One topic we want you as the reader to consider throughout the course of this book is that of ethics within engineering. The idea of ethics, morality, philosophy, and religion are inevitably intertwined, as they all consider and speak toward the idea of right and wrong. If you subscribe to this line of thinking, the concept of "ethics" dates as far back as approximately 1500 BCE, when world religions began to emerge.[2] Greek and Roman philosophers brought this concept of ethics and morality to the forefront of their research and scholarly teachings and texts around 700 BCE. Around 400 BCE, the Hippocratic oath was written, much of which is still used to this day in the graduation ceremonies of many medical schools.[3] Within today's major religions, a version of the Golden Rule exists, which in essence is nothing more than suggesting that people

---

1 https://www.theatlantic.com/science/archive/2018/05/two-spaces-after-a-period/559304/.

2 https://www.britannica.com/topic/Hinduism/The-history-of-Hinduism/.

3 https://www.britannica.com/topic/Hippocratic-oath/.

behave ethically, treating others as they would want to be treated. Beyond the overlap of ethics and religion and the concepts of right and wrong is the recognition that many ethical decisions may not present themselves in a pure right or wrong format. Ethical decisions often place an individual between the proverbial "rock and a hard place"—having to consider a variety of stakeholders and a complexity of scenarios where a given response will have a cascading effect on future decisions. At times like these it is important for an individual to consider not only their/her/his personal beliefs, but also the values of their professional society.

In 1852, the first professional engineering society in the United States, the American Society of Civil Engineers and Architects was formed in New York City, and became the American Society of Civil Engineers in 1868 when the architects formed their own professional society.[4] As the interests and expertise of engineers in the United States continued to grow and broaden, groups representing mechanical (ASME) and electrical (AIEE) engineers formed their own professional societies in 1880 and 1884 respectively. Although the establishment of these professional societies was an important step in the ultimate development of engineering ethics, at the time they were founded, ethics was considered a personal concern.[5] The concept of a code of ethics was considered by ASCE many times throughout the late 1800s and early 1900s, however, it was not until 1914 that a code of ethics was formally adopted, due in part to the adoption of a code of ethics by ASME around the same time.[6] As described in ASME's Unwritten Laws of Ethics and Change in Engineering, a number of significant structural failures in the late 1800s and early 1900s forced the entire engineering profession to confront shortcomings in technical and construction practice, which ultimately resulted in organizations requiring credentials or licensure to become a professional engineer, as well as the adoption of formal code of ethics.[5] Later in this section you will be asked to find and read the code of ethics related to your specific area of engineering. For the time being, however, we will focus on ethical codes associated with non-discipline specific entities.

Engineering in Action

Discuss with others what ethical considerations you deem important for engineers to uphold. What are the common themes or ideas amongst your group?

Two creeds or oaths worth looking at are the Engineer's Creed through the National Society of Professional Engineers,[7] and the Obligation of an Engineer from the Order of the Engineer

---

4   https://www.asce.org/about_asce/.

5   *Unwritten Laws of Ethics and Change in Engineering*, ASME Press, 2015.

6   P. A. Vesilind, "Evolution of the American Society of Civil Engineers Code of Ethics," *Journal of Professional Issues in Engineering Education and Practice*.

7   https://www.nspe.org/resources/ethics/code-ethics/engineers-creed.

ceremony.[8] While we will not duplicate each of the documents here (the footnotes/references should allow you to easily find these online), there are a number of similarities that speak to the ethical considerations that are critical to being an engineer. Specifically, each pledge speaks to the following considerations:

- The service of humanity and the consideration of public welfare
- The participation in honest enterprise, acting with integrity and fairness in all dealings
- Upholding the standards of professional conduct

Engineering in Action

Discuss again with your group the considerations listed above. What does each of them mean to you? Are there any that you previously discussed? Of those that you did not consider, do any of them surprise you?

While each of these points is an important ethical consideration within engineering, we would argue that they can in many ways be considered within a broader context of being good participants in civil discourse. While the language within the Engineer's Creed and the Obligations of an Engineer are written such that they explicitly apply to engineering, we have written them in such a way that they could just as easily apply to a physician, plumber, or schoolteacher. Let's look at these in more detail and explain how they particularly apply to engineering.

The service of humanity and the consideration of public welfare. Both documents speak directly to the idea of public good or welfare and state that engineering is a service-oriented profession. If we think about what engineers do, they solve problems in an effort to make life better for all of humanity. Think about your daily interactions and activities, and consider how many engineers helped make these interactions and activities a reality. Housing, clothing, transportation, food, communication, and power generation: all of these are a direct result of engineers from a variety of backgrounds working to help make the world a better place. The solutions we create as engineers serve humanity, and in order to do so, they need to be safe, reliable, and accessible.

The participation in honest enterprise, acting with integrity and fairness in all dealings. Here, both documents use the exact phrase "honest enterprise" in their creed, while the Obligation of an Engineer also uses the terms "integrity" and "fairness." This idea can extend in a variety of directions, including ensuring the engineer is being honest about her, his, or their abilities and calculations; ensuring the individual or company is being truthful in the source and scope of the work being presented; and ensuring that the work completed was done as intended and to an appropriate standard. In essence, an honest enterprise is one where the engineer is appropriately

---

8 https://order-of-the-engineer.org/.

qualified to do the work, the plans as presented have the best interests of the customer and public in mind, and that when completed, the work was done correctly while being economically prudent.

Upholding the standards of professional conduct. While the phrase "professional conduct" only shows up directly in the Engineer's Creed, the Obligation of an Engineer discusses upholding the "devotion to the standards and dignity of my profession." Here, both documents are not only referencing the ethical obligations of being an engineer, but they are also recognizing the importance of engineering within society, and therefore how your actions, behavior, and conduct as an engineer should appropriately represent your profession. Not only is it important to act ethically and morally in your position, but you also need to approach your job duties and responsibilities with the professionalism they deserve.

> **Engineering in Action**
>
> Look up the code of ethics for the engineering society most closely related to the engineering discipline you are studying or one that you are considering studying. Note the specific expectations associated with that society and area of engineering. Compare that code of ethics to at least one other engineering society's code. How are they the same? How are they different?

We hope that as you progress through this book and your course, you will reflect on some of these ethical considerations and how they may influence decisions you make both during your educational career and beyond.

INTRODUCTION SUMMARY

This introduction provides an overview of the textbook, along with an understanding of how each chapter will be presented. Additionally, we provide you with some thoughts regarding ethics in engineering that we hope you will consider while reading other chapters in this book, during your time in academia, and as you begin your career.

INTRODUCTION REFERENCES

Unwritten Laws of Ethics and Change in Engineering. 2015. ASME Press. Accessed December 14, 2020. http://ebooks.asmedigitalcollection.asme.org/book.aspx?doi=10.1115/1.860588

Vesilind, P. A. 1995. "Evolution of the American Society of Civil Engineers Code of Ethics." *Journal of Professional Issues in Engineering Education and Practice 121*, no. 1 (January): 4–10.

CHAPTER 1
DESIGN PROCESS

BEFORE YOU BEGIN

Take a minute to consider the following questions:

- What experience do you have with the scientific method and the engineering design process?
- When was the last time you used observations to assist in making a decision?
- What is the difference between ideation, invention, and innovation?
- What do you think of when you hear the term *entrepreneur*?

LEARNING OBJECTIVES

By the end of this chapter, you should be able to:

- Summarize the similarities and differences in the scientific method and the engineering design process
- Identify bias in observational activities
- Establish priorities for selecting potential solutions that consider existing criteria and constraints
- Develop a solution following the engineering design process
- Recognize a "wicked" problem and employ design thinking to find a solution

INTRODUCTION

Likely you were first introduced to the scientific method back in elementary school, where you were asked to conduct a simple experiment. At the same time, or perhaps later in your academic career (middle or high school), you may have been asked to develop your own question and hypothesis, ultimately reporting out your findings at your school's science fair or a similar type of event. Although the terminology and specifics of each step might differ slightly from text to text, we consider the scientific method to consist of the following steps (bolded words exemplify the typical language used when describing the steps of the scientific method):

1. Ask a **question**
2. Conduct background research
3. Develop a **hypothesis**
4. **Test** the hypothesis (**experiment**)
5. **Analyze** your results
6. Draw a **conclusion**

Some of you may have also been introduced to the engineering design process, perhaps through a high school engineering course, summer camp, or some other class or event that focused on the design or development of a product. As with the scientific method, there are different ways of describing the steps of this process, but we will consider the following as our definition of the engineering design process:

1. Identify the Problem
2. Define Criteria and Constraints
3. Gather Data
4. Brainstorm
5. Analyze Potential Solutions
6. Develop Solution and Test
7. Validate and Verify Your Solution
8. Communicate Your Findings
9. Implement Your Final Design
10. Post-Implementation Review

Much like the scientific process, the design process described here is non-linear, meaning that it is highly likely that as you walk through this process, you will need to revisit an earlier step prior to reaching the end. Additionally, both processes are iterative in nature (meaning some of these steps may be repeated over and over) and may change significantly as new information is discovered or generated. As you can see in Table 1.1, there are significant parallels between the scientific method and the engineering design process. The primary difference is that design focuses on the development of a solution to a problem. Thus, where the **scientific method** is set up to test the validity of a **hypothesis**, the **design process** is set up to test the validity of a **solution to a defined problem**. This difference also describes the fundamental difference between the *natural (basic) sciences*, which attempt to understand phenomena in the natural world, and engineering, which as an *applied science* obtains and uses knowledge to achieve a desired result (e.g., a solution to a given problem).

The scientific method tests a hypothesis, while the engineering design process validates a solution to a problem.

TABLE 1.1 Comparison of the Scientific Method and the Engineering Design Process

Scientific Method	Engineering Design Process
1. Ask a Question	1. Identify Problem
2. Conduct Background Research	2. Define criteria and constraints
	3. Gather Data
3. Develop a Hypothesis	4. Brainstorm
	5. Analyze Potential Solutions
4. Test the Hypothesis	6. Develop Solution and Test
5. Analyze Your Results	7. Validate and Verify Your Solution
	8. Communicate Your Findings
6. Draw a Conclusion	9. Implement Your Final Design
	10. Post-Implementation Review

Throughout this chapter, we will refer to the ten-step process listed here as we discuss various aspects of the design process. In addition to exploring how problems are identified by engineers and criteria and constraints are established, this chapter will focus on how your team's knowledge and experience will help in developing a solution to the problem. Additionally, sections on the topic of innovation and entrepreneurial mindset will provide extra considerations when it comes to design thinking and how these additional ways of thinking can potentially lead to a more impactful design.

Recognize that we used the word "team" when describing how knowledge and experience will help in developing a solution to the problem. As we discuss in more detail in Chapter 2, one thing that often surprises our students is how much work is done in groups or teams in engineering. Teams of engineers, scientists, and technicians are required for the development and implementation of almost any product, from Post-it notes (a team of chemists invented the adhesive and a product development team that included chemical engineers at 3M helped bring the Post-it note to market almost a decade later[1]), to the tallest building in the world (the Burj

---

1 https://ichemeblog.org/2014/11/13/a-chemical-engineer-and-the-invention-of-the-post-it-note-day-163/

Khalifa, which stands at 2716.5 feet, took 22 million man hours to build, and included over 60 different consulting companies and 30 on-site contracting companies from around the world[2]), to sending men to the moon (the Apollo program employed over 400,000 engineers, scientists and technicians during its 14 year period[3]). While this list may appear to be extreme examples of the teamwork necessary to develop a product, engineers with a variety of experiences and expertise will be necessary over the lifetime of the design process for all products and projects. Although the titles and expertise may vary from discipline to discipline, the teamwork required will not. As an example, in the development of a medical device, the project manager may be working with design engineers, manufacturing engineers, regulatory and legal experts, production and systems engineers, quality assurance and validation engineers, transfer engineers, packaging and environmental engineers, research and development and more! This doesn't even begin to consider the fact that those design engineers might have individual sub-systems they are working on, including electrical, chemical, mechanical, software and more. Teaming is a crucial part of engineering and being able to work together as a team throughout the design process will be critical in the success of your final design.

PROBLEM IDENTIFICATION–OBSERVATIONS

Step one in the engineering design process is to identify a problem to be solved. In the world of product development, this is often referred to as *opportunity identification*, where opportunity is defined as an idea for a new product. One question that often arises from students as they begin the design process is, "How do you identify a problem?" In many design classes, problems have already been identified by the instructor to ensure their scope and complexity are appropriate for the course. Furthermore, really understanding that the problem you are intending to solve is the "right problem" can take a lot of time and effort, perhaps more than the rest of the design process! As a professor of industrial design at Yale University once said, "If I had only one hour to solve a problem, I would spend up to two-thirds of that hour in attempting to define what the problem is."[4] This level of focus on the details of the problem itself is also evident in the root cause analysis undertaken by the Toyota Production System, where using the "five whys" method "by repeating 'why' five times, the nature of the problem as well as its solution becomes clear."[5] Many times, particularly in manufacturing, one will find that this type of analysis allows the team to trace a problem back to its origin, revealing that the actual problem that needs to be addressed may not be what was originally thought to be the problem (*root cause analysis*). For

---

2 https://www.burjkhalifa.ae/img/FACT-SHEET.pdf
3 J. O'Heir, 'Apollo Remembered', *Mechanical Engineering*, vol. 141, no. 07, pp. 28–33, Jul. 2019, doi: 10.1115/1.2019-JUL1.
4 Robert E. Finley, *Manufacturing Man and His Job* (American Management Association, Inc.: 1966).
5 T. Ohno, *Toyota Production System: Beyond Large-Scale Production*. (CRC Press: 1988), 176.

example, you might choose to take acetaminophen (Tylenol) or ibuprofen (Advil) if you have a headache, but the underlying cause of that headache (lack of sleep, dehydration, etc.) will likely not be addressed by the painkiller. While in the short term the analgesic does its job (i.e., you no longer have a headache), it is quite probable that the headache will return once the drug wears off unless you adequately address the root cause of your pain.

> ### Engineering in Action
>
> Discuss with one of your classmates a problem or issue you've recently encountered. Allow them to form "why" questions to determine the root cause of this problem or issue. For example, perhaps your problem or issue is that you were recently late for work. The first "why" question might reveal that you were late for work because you left your house later than you should have. Further probing might reveal that by going to sleep earlier, perhaps you wouldn't be late for work. Once you've determined the root cause of your problem or issue, switch roles and repeat.

Problem identification can originate from many different inputs, including the dissatisfaction of currently available solutions, an overall lack of solutions for a given problem, or technology advances that allow for new solutions to a given problem. How do you know a problem exists, particularly a problem worth solving? There are many methods one can use to find these pain points, including interviewing or surveying potential beneficiaries of the new product or technology, typically clients or end-users of the product; looking at reviews of products that already exist on the market, particularly those that describe limitations with the current solution; or observing either the situation that warrants a solution or how the end-user interacts with current solutions. The idea of making *observations* certainly fits within the problem identification step of the engineering design process and could either preclude or be a part of asking a question in the scientific method. Below we present six things to think about as you make your observations; many of these revolve around the idea of trying to eliminate preconceived notions about the observation, in other words, entering the situation with an open mind.

Avoid Bias

Bias in relationship to observations (sometimes referred to as experimenter or research bias) focuses on the fact that **as an observer, we tend to see what we expect or want to see**. Preconceived ideas or expectations can prevent us from seeing what exactly is going on or may slant our analysis in support of our prior expectations. Often this bias can be *implicit* or unconscious, meaning that even though you are not consciously or explicitly trying to be biased, the decisions you make are based in part on what makes you uniquely you, such as your previous experiences, family, race, and gender. There are many examples of this type of bias in product development. For example, the fact that crash test dummies historically were modeled after "average" male heights and weights means that seatbelts were not appropriately designed for women, making female drivers 73% more likely to be seriously injured in a car crash. Automobile manufacturers did not explicitly decide to make their cars more dangerous for women: it just so happened that back in the 1960s and early 1970s, the first crash tests (which ultimately resulted in the development of the first crash test dummy in 1976) were performed on men![6]

Engineering in Action

You are working for a company that is looking to enter the wearable electronics market to aid in fall detection. You have set up a consumer focus group to observe a group of consumers (all over the age of 65) using several competing devices currently on the market. You observe that several of the consumers have a difficult time interacting with the device and become visibly frustrated when trying to use it. From your own personal experience, you've observed that your grandfather, who happens to be of similar age to the consumer group, avoids using electronic devices, as he finds most of them too technically difficult to use. As a result of this bias, you conclude that the current devices on the market are too technical to be used appropriately by the target audience, and as such, your product would need to be simple to operate.

While this might be a reasonable conclusion, what other potential reasons might explain the difficulty the consumer group had in using the other products?

Possible explanations: The size of the device might cause issues with consumers who have limited visual acuity and/or reduced manual dexterity.

View Multiple Times

Like any experiment you conduct, any observation you make should be done multiple times to verify that what you observed is an accurate representation of the event in question. For example, suppose you go to see your favorite band perform live, and the lead singer is not having a good night. You might conclude that they are not a great singer and that their studio albums are obviously all heavily engineered. If you instead go to four or five concerts, and the singer is

---

6 Keith Barry, "The Crash Test Bias: How Male-Focused Testing Puts Female Drivers at Risk," *Consumer Reports*, October 23, 2019, https://www.consumerreports.org/car-safety/crash-test-bias-how-male-focused-testing-puts-female-drivers-at-risk.

fantastic in those performances, you might conclude that your first observation was an anomaly, and in fact, they are a great singer.

Make Naturalistic Observations

Social scientists refer to naturalistic observations in as much as they are observing behaviors in the environment in which they typically occur. If you want to see how kids interact in a classroom, you would not observe their interactions during lunch or recess; you would want to do so in the natural environment (the classroom). Furthermore, you would want to minimize your interactions with the children by observing them from behind a mirrored partition or through a camera, for example. Zoologists use the same principles for observing animals and animal behavior. Obviously, if you're developing a new kitchen device, it would be better to observe people's behavior in the kitchen versus in their office to determine what needs they might have for the new device.

Follow Up with Questions/Clarification If Possible

Going hand in hand with what we previously described in the section discussing bias, following up observations with questions to clarify some of the things observed is an excellent way to ensure the validity of the conclusions you have drawn because of your observations. Obviously, you want to make sure any question(s) you ask help you better understand what it is you want to learn; to do this, you want to make sure you have adequately prepared your questions ahead of time and thought through different scenarios such that you can develop appropriate follow-up questions based on a variety of anticipated responses. While you cannot anticipate every answer (since you are doing these observations to obtain information and gain knowledge), the more you can plan ahead and have a series of questions that focus on what you want to learn, the better the interaction with the user will be.

Engineering in Action

From our previous example of the product design of a fall sensor, what additional questions would you want to ask your focus group to better understand the limitations associated with their interactions? Make sure your questions avoid leading the focus group to a particular answer and that they are worded in a way that limits "yes" or "no" answers.

Be Nonjudgmental

When you are observing someone or a group of individuals, you might be trying to understand what limitations are present in existing products designed to solve a specific problem, or you might be trying to observe what people naturally do when confronted with a problem that currently has no solution readily available. In doing this, you are trying to get additional insight into what opportunities exist for product innovation. You might feel the need to question why someone is behaving the way that they are; in your opinion, their behavior may be nonsensical based on your knowledge of the problem or available solutions. However, it is critical to remember that your goal is simply to collect information based on your observations such that you have the appropriate information to make an informed decision. Because collection is the goal, it is important not to begin to interpret, analyze, or judge the information at this stage.

Avoid Trying to Solve the Problem Immediately

This is probably one of the most difficult things to avoid doing when one begins an observation. Our inclination as engineers and as people is to immediately attempt to solve the problem we are seeing. If you are following the engineering design process previously outlined in Table 1.1, unless you have already identified a solution and are using observations as a means of testing and/or validation, you are likely in one of the first three phases of the design process (identifying problems, defining constraints and/or collecting data). At this point in the process, again, your observations are being done to collect unbiased, non-judgmental information. Attempting to solve the problem at this stage might cause you to do so without enough information to truly understand its root cause. Attack a problem just as a detective such as Sherlock Holmes might try to solve a mystery. Question everything, leave no stone unturned, and consider all angles and possibilities when looking at potential solutions.

Attack a problem by questioning everything, leaving no stone unturned, and considering all angles and possibilities when looking at potential solutions.

DEFINING CRITERIA AND CONSTRAINTS

After you have identified the problem that needs to be solved, you need to understand what the limitations of your design can or must be. For example, how long do you have to bring your design to fruition? (Although we often will refer to the end result of your design as a product, recognize that this could just as easily be a plan for a new construction site, dimensions of a structure, loading rates for a treatment process, or even steps for a more optimized process.) What is your team's area(s) of expertise? What is the development budget for your project? Beyond the constraints you might face organizationally, there are specific criteria that need to be met for the design to be deemed "successful". Oftentimes this is something that is translated from what the user (client) says the device or design needs to be able to do. This in and of itself can be a tricky endeavor and one in which you will likely use many of the suggestions we described in the previous section regarding observations. While in the end the customer is "always right," their proximity to the problem and perhaps their lack of knowledge of possible solutions may potentially lead you toward developing solutions that do not really address the real problem or reduce the impact your design could have. Henry Ford is often quoted (albeit this quote is unsubstantiated[7]) as having said, "If I had asked people what they wanted, they would have said faster horses."

To ensure needs are actionable by the designers and engineers, you will need to translate them into tangible engineering criteria to better define the array of possible solutions. As a simple example, let's imagine a client comes to you to design a new eraser for their pencils, as their customers have told them they are tired of how their current pencil erasers leave fragments of the eraser on their work surface when they are used. If the user need is to prevent the eraser from leaving debris on the work surface, perhaps the design criteria or input is that the material the eraser is made from will minimize the amount of dust it creates or is formulated such that the dust accumulates into larger fragments that can be easily removed from the work surface. In this case, the user need is translated into a design input or constraint that focuses on specific types of materials or chemistries to be satisfied. Going through other customer needs can result in additional criteria such as size, color, hardness, and so on.

Engineering in Action

Break off into groups of three or four. Spend a few minutes individually thinking about and writing down what you might want as a customer in a new pencil eraser. Share these "needs" with your group. Then, as a group, pick five or six of the best needs for further discussion. From there, determine how you might turn those needs into engineering criteria (size, weight, shape, material, cost, etc.).

Overall, we can think of criteria and constraints as our *engineering or design inputs*, as they define what our end solution must accomplish. As we will see later in this chapter, it is quite possible that our design inputs will conflict with one another, and as a result, we might need to make compromises or prioritize our requirements. In order to fully understand your design parameters based on these inputs, you may need to have a description of the input as well as appropriate values associated with it. Going back to the activity we asked you to complete regarding the pencil erasers, one of the possible criteria suggested was size. Did you define a size? How big should a pencil eraser be? How big is your current pencil eraser? To have this information available when you consider possible design options, you will probably need to gather additional data prior to proposing possible solutions.

---

7 Patrick Vlaskovits, "Henry Ford, Innovation, and That 'Faster Horse' Quote," August 29, 2011, https://hbr.org/2011/08/henry-ford-never-said-the-fast.

GATHERING DATA

You'll notice that we are several pages into our chapter on design, and yet we still haven't discussed the physical action of designing something. It's natural to want to jump right in and begin with a solution, similar to how we often want to make judgments and solve the problem when making observations. **But it is important to see what alternatives already exist that may serve as possible solutions to our problem.** Not only does this prevent us from re-creating the wheel, it can also serve to generate ideas both on what can work and possibly even what has been tried in the past that clearly did not work. Gathering data allows us to see both the historical context as well as the current landscape in which our problem is situated. While some information might be gathered by observing specific behaviors and phenomena as previously described, such as determining how a consumer interacts with a product or how many cars pass by a potential retail site each day, other data will need to be researched and gleaned from a variety of external sources. Depending on our familiarity with the problem, the data we gather may be from internal sources (e.g., looking back at a similar project completed last year, creating a new version of one of our current products, or considering the current process) or primarily from external sources (e.g., competitors' products, existing designs, or established practices).

Chapter 4 of this textbook looks at *information literacy*, and many of the suggestions and recommendations in that chapter focus on the authenticity and authority of your resources. Although reading that chapter right now may give you greater insight into why authenticity and authority are important and how one can assess the validity of a chosen resource, the messaging that is important at this time is to understand where the information you are gathering is coming from and to what level you trust that information. If you are building a seventy-five-story luxury apartment building, you probably want to make sure that the specifications and regulations you researched are correct. In other words, you want to make sure you have validated and verified your data sources.

Target values for your design inputs can come from a variety of sources, including your own observations; previous or current designs sourced internally or externally (i.e., designs from your own company or designs from competitors or competing products); or they may even be functions of limitations or requirements set forth based on the project, your company, or local, regional or federal requirements. For considerations such as size, your targets might be based on a variety of factors depending on what it is you were designing. A few examples of these targets are listed below.

- Designing a handheld vacuum: you need to consider the limits of human performance when determining the weight of the object
- Designing a hip implant: you need to consider the average geometry and size of the acetabulum (hip bone socket) and diameter of the femur (upper leg bone) in a human

- Designing a new building: you need to consider local regulations as well as the proposed site for the building
- Designing a new air handler for a building: you need to consider regulations, the size of the space, and the number of people that will be in the building

> **Engineering in Action**
>
> Return to your groups of three or four. Spend a few minutes looking over your pencil eraser criteria. What reliable sources might you examine to determine target values for your criteria? Attempt to develop target values for your criteria, making sure to cite appropriate sources as necessary.

One source of data that you will need to consider in just about any engineering field is the regulatory requirements regarding your design. This may have different connotations for different engineering disciplines. In designing a building or building site, understanding local, state, and/or federal codes for construction requirements may be important. For example, a structure in California may require reinforcements to account for loads associated with earthquakes, while the same structure in Florida may be required to withstand specific wind loads to resist hurricanes. Because these local considerations result in different design inputs, the final design might look very different, even if the functionality of the building in those two areas are the same. Similarly, for biomedical engineers, the Food and Drug Administration (FDA) has specific requirements for different types of medical devices depending on how life-threatening that device's failure would be. These regulatory agencies provide another layer of requirements a design engineer must be familiar with when determining appropriate design inputs.

MAKING AND JUSTIFYING ASSUMPTIONS

Throughout the design process, design engineers qualify their designs with specific criteria and constraints to better define the scope of the problem to be solved. **These criteria and constraints can arise from user inputs, regulatory requirements, local laws, and other governmental or societal expectations.** Sometimes, however, there may not be appropriate data to include as design inputs. You might be tasked to solve a problem where there are no competing products on the market; as such, using other products as benchmarks to develop target values may not be possible. You also might find that even if there are similar products in the marketspace, some companies choose not to publish their product specifications; therefore, obtaining that information might be difficult. Alternatively, you might be tasked with designing for a specific task, geometry, clientele, or space requirement where more common designs with similar functionality may not be appropriate as benchmarks. Lastly, you might be trying to design something so complex that even with well-defined criteria and target values, the problem at hand is still too complex to solve. In these cases, what is an engineer to do?

> **Engineering in Action**
>
> Consider the use of stairs as a potential solution to allow for access to the second level of a structure. What assumptions are associated with this option as a viable solution? How might the following criteria, constraints, additional information impact your assumptions: the need to consider individuals with limited mobility, a lack of power, and the fact the structure is a theatre that is part of a park and is designated as a national historical landmark rather than a typical building (Colorado's Red Rocks Park).

One method engineers commonly use in developing target values for specifications or design inputs is the development of *assumptions*. While in many circumstances the concept of an assumption is viewed as being negative (you have probably heard the negative saying about what happens when you assume), **in engineering, assumptions are commonly used to help constrain or simplify the problem to a point where it can be solved**. By definition, an assumption is described as something that is accepted to be true, with the caveat that it is accepted without question or proof. The last part of this definition is where the negative connotation associated with assumptions arises. In engineering, assumptions are accepted to be true, but not necessarily without proof. Instead, assumptions are built from past experiences, either through your own personal history or that of the design team, company, or industry. Many of the codes and regulations that have been developed over the years when it comes to building specifications, material properties, regulatory requirements, and so on are derived from previous experiences and/or experimentation, which now is provided as written evidence (and as previously described additional criteria) of past experiences.

While these assumptions are often made early in the design process to simplify or constrain what is often an open-ended problem, many times these assumptions also result in permanent restrictions on the solution, such as weight restrictions on products, seating capacities in buildings, or even speed limits on roadways. While these limits are always presented to the end user, it is also important to recognize that there may be cases where these restrictions aren't followed. For that reason, it is common to include a *factor of safety* on values associated with these restrictions. Factors of safety are ratios that compare actual failure values to recommended maximum limits. A ladder may have a maximum recommended weight limit of 300 pounds but is unlikely to completely collapse if someone weighing 301 pounds climbs it. The ladder itself might be able to hold 330 pounds, but to improve safety (and reduce the chance of injury and potential litigation) the manufacturer selects to set the limit at 300 pounds. The ratio of 330/300 = 1.1 is the factor of safety for which the ladder is designed. This design has several assumptions factored in. First, there is the assumption that any individual using this ladder is aware, at least to a degree of accuracy, of how much they weigh. Second, individuals using the ladder recognize that the weight limit is a total weight limit for the ladder; in other words, two individuals using the ladder cannot have a combined weight of more than 300 pounds. Third, the added cost of manufacturing a ladder that can hold 330 pounds and yet only be advertised for 300 pounds is "worth" the increased safety. Last, and perhaps most important, these factors of safety are often not an arbitrary choice of the design engineers. They are typically a result of multiple factors, including regulatory code and enforcement, tolerances in materials and manufacturing techniques, and the impact of a potential failure, just to name a few.

Engineering in Action

Consider an elevator with a posted maximum load of 1500 kg or 15 people. Take two minutes to come up with a list of assumptions related to this restriction. Once you have completed your list, compare it with a classmate's.

You will notice that making and justifying assumptions is not an explicit step in the overall design process. In fact, you'll find that engineers use assumptions throughout the design process, all the way from defining criteria and constraints (as we have described here) to solution analysis as well as throughout the testing and validation process. As you progress through your engineering education and your career, you will become more adept to understanding how and when the use of assumptions is justified in your daily operations. You will recognize that many of the regulatory restrictions we follow in designing a vast majority of our projects are based in part on engineering assumptions that have been validated, or perhaps revised over the years through direct observation, experimentation, and engineering analysis.

BRAINSTORMING

Once you have completed your observations, defined your criteria and constraints, collected the relevant data, and made any necessary assumptions, it is time to begin brainstorming. As with many of the earlier steps, brainstorming can be less effective, or even become fully derailed, if we do not treat it as completely separate from subsequent steps. **Brainstorming is the generation of as many ideas as possible that can be considered potential solutions to the stated problem.**

While we want to take into account the prior steps in the design process, we also want to make sure that during this stage we are focused on idea generation and not on analysis, judgement, or discussion.

Brainstorming originated back in the 1940s and early 1950s; the term itself was first introduced by Alex Faickney Osborn in the early 1950s.[8] There are many different methods one can use to brainstorm, but most of them work best when a facilitator is involved that can keep the group on task. As you figure out what methods work best for you and your team, consider the following guidelines that are generally accepted "best practices" for brainstorming. Note that some of these come from Osborn's rules, established almost seventy years ago.

1. **Be non-judgmental.** As we suggested earlier, brainstorming should be focused on idea generation, not analysis or judgement. In addition to verbal judgment, make sure you are also being non-judgmental in your posture, expression, or other nonverbal means of communication.
2. **Build upon the ideas of others.** Allow ideas that have already been presented influence your own thinking. Doing this not only allows for the generation of many different ideas (the goal here is quantity) but helps to create a positive and inclusive environment, which in and of itself will encourage additional participation.
3. **Encourage wild or unconventional ideas.** It is always easier to tone down an idea during the analysis phase of the design process versus attempting to restrict yourself to only conventional solutions during a brainstorming session.
4. **Encourage quantity over quality.** This has been alluded to in the first three suggestions on this list but is important enough to warrant its own suggestion. At the end of the session, you will be tasked with going back and looking over all the ideas (which means you'll want to be writing these down; handing out pads of sticky notes to each participant is a great way of doing this), so quality can be assessed at a later time point.
5. **One idea at a time.** A good facilitator will make sure everyone in the group participates, but that this is done in a way where each person has an opportunity to speak and be heard by the rest of the group. While these sessions may appear chaotic, many of the other "rules" listed here cannot be accomplished unless the group is participating both in the generation of ideas as well as listening (but not judging!) to everyone else. Although many of the descriptions we've used here suggest verbal communication, we recognize that drawings might be the best means of communicating your idea to the group, particularly for physical designs, so don't feel you need to only be writing during this process.
6. **Set a time limit.** Setting an appropriate time limit (most experts suggest 15–45 minutes) prior to starting can help make sure you stick to the task at hand and are being productive. Times that are too short or too long may not give you time to generate enough ideas or may leave you being less productive at the end of the session. The amount of time you spend

---

8   Alex F. Osborn, *Applied Imagination: Principles and Procedures of Creative Thinking*, (New York: Scribner, 1953), 317.

on this activity likely will depend on the complexity of the identified problem, your team's experience with brainstorming, and other factors external to the project (such as how many minutes you have left in the class period!).

ANALYZING POTENTIAL SOLUTIONS

Now that our brainstorming session has generated a comprehensive list of potential solutions, we can begin to analyze each of these solutions. Many times, people begin this analysis by cycling through the list and condensing similar ideas. An additional step would be to identify any ideas that may not meet the criteria or constraints previously defined (during a productive brainstorming activity, it's possible a few of these are inadvertently included, which isn't an issue and is something that is easily fixed). Once a revised list is created, the next step is a more in-depth analysis of the solutions. For this step, it is often necessary to create not just a binary "meets" or "does not meet" option for the various criteria and constraints but also a more developed prioritization of these categories and a more refined scale for how each one is met. Establishing these priorities before selecting desired solutions ensures that the selection is based on a standardized set of expectations that can be reasonably replicated rather than a selection process that may include shifting standards or unintended bias.

Establishing standardized expectations reduces the chance of shifting standards or unintended bias being included in the selection process.

One way to develop a selection process that helps mitigate standards shifting or unintended bias is to look back at the research you did for your design inputs and determine how well products already in the marketspace address your design inputs. If there is a product that is a market leader (recognizing that sales and quality do not always positively correlate) or an aspirational product you would like your solution to eclipse, you can use that as a benchmark against which you rate each of your ideas. For every design input, you can rate your potential solution to be better than (+1), equal to (0) or worse than (-1) your aspirational product(s). In this type of analysis, you might also consider adding weights to your design inputs, recognizing that some of them might be more important than others, and as such should be counted more heavily in your tally. Tallying the scores for each potential solution (and accounting for weights as appropriate), should allow you to objectively rank your ideas based on your perception of how well each idea meets your design requirements. Depending on the number of potential solutions, you may need to complete additional rounds of this activity, perhaps increasing the sensitivity of your scale. For example, instead of a three-point scale, you increase the scale to five points, providing additional means of differentiating performance that is better than the benchmark (four or five points) versus performance that is lower than the benchmark (two or one point/s). After doing this, you will likely have one or more ideas that you want to consider pursuing further.

As an example, let's return to our previous example of a pencil eraser. Through your research, you have established that the gold standard for erasers (aspirational product) are those flat pink rubber erasers that have the beveled edges. You also establish that the most commonly used eraser (market leader) is the one that is on the back of a pencil, which also happens to be made of rubber. You have formulated two new compounds for making erasers, one that does not create any dust and one that creates larger clumps of dust such that they are easier to remove from the paper. Based on your design inputs, your analysis might look something like Table 1.2.

TABLE 1.2 Comparison and Analysis of Potential Design Options for New Eraser Materials

Design Input	Pink Bar Eraser*	Pencil Eraser	Compound A	Compound B
Complete removal of marks (20%)	0	0	(-)	(+)
Minimal damage to paper (20%)	0	(-)	(-)	0
Amount of debris/dust (35%)	0	0	(+)	(+)
Cost (20%)	0	(-)	0	(-)
Formability (5%)	0	(+)	(+)	(-)
Total	**0**	**-0.35**	**0**	**0.3**

*Indicates product against which ratings were made (product standard). (-) indicates product or concept is worse than, 0 indicates product or concept is equal to, and (+) indicates product or concept is better than the product standard. Compound A is a rigid substance that minimizes dust while Compound B is a softer material that creates large clumps of dust for easier cleaning.

Note that when we compared our ideas, we used the design inputs we previously generated as comparison points in this table. For this example, we also weighted our design inputs, providing the greatest weight to the design input that was the most important (amount of debris or dust generated when using the eraser) and the least weight to the input that we determined to be the least important (formability). We then compared our two different ideas as well as another common market product to our standard and assessed (in the case of the pencil eraser) or estimated (in the case of our compounds) their performance relative to that standard (these predictions might be based on the chemical or mechanical properties of the compounds being used, or you might also have some preliminary data on your potential design solutions that you could use as part of this analysis). As you can see in Table 1.2, both compounds are predicted to be better than the pink bar eraser in two of the design input categories, equal to the pink bar eraser in one of the categories, and worse than the standard in two of the design input categories. If we had not weighted our inputs, both Compound A and Compound B would be comparable to our standard. However, by assigning a level of importance (weights) to each input, we found that Compound A, while it was very good at minimizing (actually eliminating) the formation of eraser dust, did not rank as highly as Compound B. If the weights of the inputs were to change (if, for example, formability or cost were more important), these results might change.

Before we get to testing and validation of your potential solution, one additional piece of analysis may need to be considered. In your gathering of data, you likely used competing products to aid in the determination of your design input target values. Furthermore, we just suggested using a product already on the market as your benchmark standard and a means of analyzing and ranking potential solutions. Beyond what is actually available in the market, you may also need to find evidence of what might be coming to market or what other companies have considered and potentially protected. To do this, you will likely have to search for *prior art*, that is, evidence as to how others have attempted to solve the same or similar problem. As a definition, **prior art is any information previously released that contains evidence as to how an invention works**. While prior art can be found in just about any type of publication, including books, trade journals, and company press releases just to name a few, the term is most often used when one is referring to a *patent*. A patent is a license granted by the government that gives the owner exclusive rights to make, use, and sell the invention described in the patent for a specified period of time, typically twenty years in the United States. By granting this exclusive right, the patent owner by definition has exclusive rights to forbid anyone else from making, using or selling that invention. Even if you did not know a specific patent existed, if your idea or design infringes upon someone else's intellectual property (a patent describes the intellectual property) they have the right to prevent you from making, selling or using that invention while their patent is still active. Generally, knowing what the patent landscape looks like prior to testing and validating your possible solution is a valuable way of not spending too much time developing a solution that is already protected.

Prior to testing, research the patent landscape so you don't spend too much time developing a solution that is already protected.

So, you may ask yourself, why would a country or government want to issue patents? If you want the best ideas, the best designs to be developed and manufactured, what benefit is it to society if a patent is granted but the inventor never makes the invention? Why should we grant someone exclusive rights for a period of time? The answer, perhaps non-intuitively, is to encourage additional innovation. The easy one to see, and a big motivator for companies to file patents is the fact that your design will be protected for a period of time. That protection might give your company a competitive advantage over your rivals. Although not all types of engineering firms utilize the patent process, some, such as those who work with medical devices and/or computer technologies, patent a lot of their ideas and inventions as a means of protecting their intellectual assets and giving their company that competitive advantage. The other way innovation is encouraged is that a patent must tell people exactly how your invention is made. By providing this information, you are in essence giving your competition the keys to how your invention works and encouraging them to figure out a way to work around your design by developing something new and novel. If a company can figure out how to build a design without violating your intellectual property, they are free and clear to do business.

> **Engineering in Action**
>
> Go to the United States Patent and Trademark Office website (uspto.gov) and search for any patents related to your favorite electronic device. How many did you find? When was the first one granted? How about the most recent?

TESTING AND VALIDATING YOUR SOLUTION

Once we have selected our preferred solution (or sometimes solutions), and we have validated that the method of our solution does not violate any intellectual property, we need to check to ensure that what we think will work actually is a solution to our problem. Not only that, we also need to make sure that it is a viable and competitive solution. This can be done, in part, by looking at how the proposed solution aligns with the criteria and constraints of the problem, considering what assumptions are included in the proposed design, and confirming that the solution addresses the most critical priorities of the problem. While these steps are important, testing and validation goes beyond these routine checks.

For a majority of our students, testing and validation of a solution focuses on rapid prototyping using CAD and 3D printing. While these tools have revolutionized the prototyping stages of design in many engineering fields, we want you to recognize that testing can take many different forms, and ultimately is a way of ensuring a solution can actually do what we think it can. Testing might include CAD drawings to better understand how each component in a design interacts with the others, as well as an understanding of how well the design's size and shape fit within our design parameters. Alternatively, crude mockups of off-the-shelf parts might be used to validate not only the basic size of the proposed structure, but also to better understand how a user might interface with the design. More detailed physical models (full scale or partial scale) can also be employed to better understand how the design will be physically represented in the three-dimensional space. While some of this modeling can be done by additive manufacturing techniques (3D printing), subtractive techniques (cutting, milling or other standard machining techniques) on rapid prototyping materials such as foam, paper, wood, clay, or other physical

CHAPTER 1 DESIGN PROCESS 17

media may be easier and quicker to accomplish. Being able to quickly develop these models will allow you to potentially go through multiple iterations and revisions to your solution in a short period of time, allowing you to understand if you are trending toward a viable solution or are demonstrating that your proposed solution does not work as you had originally envisioned. **While there are times solutions can be revised to the point of success, sometimes it may be easier to look at other promising solutions on your original list to see if any of those can be further developed.**

Many other forms of testing and validation exist beyond what we have previously described, which could be thought of as feasibility or ergonomic testing. Mechanical or finite element (mathematical) testing of materials, joints, or other parts of the design may need to be accomplished to verify that wear or other types of mechanical failure do not occur. More careful ergonomic testing may need to be done to fine tune interaction points such as triggers, switches, knobs, and touchscreens to ensure ease of use and robustness of design. In designing systems that include electronic hardware or software, logic testing and fault testing may be necessary to ensure your design works appropriately. Any part of your design you need to verify or validate should be tested such that you can confidently say, and appropriately document, that your design will work the way it was intended. In a general engineering book such as this, listing the different types of tests across all the disciplines would be a monumental task; later in this section we will mention two specific organizations that play an important role in the development of standardized testing protocols used worldwide. You will likely encounter these and other, more discipline-specific and perhaps even regionally-specific ones as you progress through your degree and engineering career.

One thing to carefully consider in any test you conduct is the methodology of your testing protocols. You have likely completed many different tests or experiments in your educational career: perhaps things like acid–base titrations in chemistry class or a frog dissection in biology class. Each of these experiments (tests) had step-by-step instructions as to how to complete the experiment. But how did the person writing the lab know what the molarities of the solutions should be to start with or that adding the buffer dropwise would achieve the desired result? Similarly, if you were to run a mechanical tensile test on a piece of metal, how would you know what cross-sectional area to use, or how quickly you should elongate the material? (A tensile test works by lengthening a material of a known cross-sectional area by a specified amount per time [loading rate, usually in mm/min] and measuring the force [load] needed to lengthen or displace it.) Does it depend on the material or what the material is supposed to do in your design? How do we arrive at the various parameters for a test?

If you run that tensile test often enough, you might ultimately be able to determine the correct parameters that allow you to translate your testing results to real-world performance simply by serendipity. Obviously, this method is time-consuming and is not an ideal way to determine these parameters. A better way would be to gather additional data and find evidence from documents such as journal articles, academic textbooks, and trade and governmental handbooks to better understand how other people have conducted these types of tests. One potential issue you may come across is that different groups use different methodologies in their tensile testing; perhaps each group uses a different loading rate and cross-sectional area. While this is certainly better than the infinite possibilities we have without doing any research, you may find for many applications, particularly in regulated industries such as buildings and medical devices, that the parameters have already been established by *standards* that the regulatory bodies require designs be tested against for validation and verification purposes.

Although there are a number of international standards organizations that cover a variety of areas that affect your daily life, including electrical and electronic technologies (the International Electrotechnical Commission or IEC), telecommunications (International Telecommunication Union or ITU), and air conditioning (American Society of Heating, Refrigerating and Air-Conditioning Engineers or ASHRAE), there are two standards organizations that are frequently referenced when it comes to the development of physical testing protocols; that being the International Organization for Standardization (ISO) and ASTM International (formally known as the American Society for Testing and Materials). These organizations produce established requirements for a variety of technical tasks to ensure that testing across organizations and disciplines is done in a way that is repeatable and verifiable. These standards are developed by groups of individuals who are experts in the standard's area of interest, including representatives from the various companies that would benefit from the establishment of such a document. These standards are then referred to by regulatory bodies to ensure that new designs meet specified minimum standards and allow for direct comparisons between different products or materials. Depending on the type of device you are developing, certain specific testing standards may be required, while for other types of devices, you may choose a specific standard based on the type of test you need to complete. In both cases, using standards whenever possible ensures you are meeting specific criteria that has been established by academic and industrial experts in the field.

COMMUNICATE YOUR FINDINGS

Although we list communication as step eight in the design process, and we specifically focus on communicating your findings, good communication is something that occurs throughout the engineering process. We will discuss written communication in detail in Chapter 5 and for the communication of your findings, the sections in Chapter 5 covering technical reports and laboratory notebooks are highly pertinent. Although we won't go into the specifics regarding communication here in this chapter, one thing we will reiterate here is a quote from Tom Clancy: "If you don't write it down, it never happened."[9]

---

9   Tom Clancy, *Debt of Honor*, (New York: Penguin, 1995), 108.

INNOVATION

So far, we have spent a good amount of time talking about the design process. If we have appropriately worked through the ten steps of our process (actually, at step nine), we have developed a solution to our problem. Is what we have developed an invention? Can we design something without inventing something? In addition to terms we have already used such as design, you may have heard other terms thrown around when describing the design process, words like "ideation," "invention," and "innovation." Can these be used interchangeably, as they often are?

Before we get into answering these questions, let's discuss another way of looking at solving problems, specifically the idea of *design thinking*. Touted as a methodology to solve "wicked" problems (those that are ill defined or unknown), design thinking includes five phases that correlate well with the engineering design process (as shown in Table 1.3). It is a method of "thinking outside the box," designed to develop solutions that might not be readily apparent and encourage innovation. **The design thinking steps are typically thought of as phases, since, even more so than the engineering design process, they often occur non-sequentially and repeat iteratively.** This process is also meant to work such that the ideation, prototyping, and testing phases occur rapidly; thus, many prototypes are tested to determine what works best and what continues to need improvement.

TABLE 1.3 Comparison of the Engineering Design Process and Design Thinking

Engineering Design Process	Design Thinking
1. Identify Problem	1. Empathize
2. Define criteria and constraints	2. Define
3. Gather Data	
4. Brainstorm	3. Ideate
5. Analyze Potential Solutions	4. Prototype
6. Develop Solution and Test	4. Prototype
7. Validate and Verify Your Solution	5. Test
8. Communicate Your Findings	
9. Implement Your Final Design	
10. Post-Implementation Review	

The key difference between the engineering design process and design thinking is that design thinking focuses on *empathy*; not only identifying the problem but gaining an intimate understanding of the issues associated with the problem from the viewpoint of the potential beneficiaries of your design. Making humanistic decisions in developing a solution is an important aspect of the design thinking process. The ability to relate and truly understand the issues at hand will help you develop a more meaningful solution and allow you to completely understand the benefits and costs associated with design decisions and changes.

Design thinking includes empathy, the intimate understanding of the issues at hand from the viewpoint of the beneficiaries of your design.

Ultimately it is the concept of empathy that allows problem solutions to turn into innovation. All of the other terminologies across the two proposed methods have appropriate analogs. For example, within the phases of design thinking, we introduce the word *ideate* and equate it to brainstorming in the engineering design process. If you dig deeper into the design thinking paradigm, brainstorming is defined as a method of ideation, with other popular methods including braindumping (similar to brainstorming but done as an individual), brainwriting (similar to brainstorming, but ideas are written down rather than explained discussed) and brainwalking (brainwriting done at various locations throughout the room, with participants moving from location to location to add to the ideas generated at that location). For the purposes of this text, we will consider ideation and brainstorming to be one and the same, just as we would feel comfortable using the terms *prototype* and *develop solutions* interchangeably.

Empathy, on the other hand, has no direct correlation in the engineering design process, and it is within the constituents of empathy that we begin to see the nuances between design solutions and innovations. The engineering design process is centered around the concept of solving problems, which oftentimes (but not always) will have a perceived benefit for a target audience, while innovation is centered around the person for whom the invention will benefit. In many cases, these two processes are likely to converge at the development of a solution that meets the needs of an end user or consumer, but that is not always the case. The engineering design process can just as easily develop a solution that, while it solves the problem at hand, due to the nature of the solution may not be practically beneficial to the end user or consumer. As such, while they solve the problem defined by the design team, these solutions do not rise to the definition of an innovation, as they do not ultimately improve the market or societal target from which the initial problem arose. An example might be the development of a new car that gets one thousand miles to a gallon of fuel, but the fuel is so unstable that its use is only practical in an extremely controlled laboratory environment. While this is an engineering solution, and while the development of the car or fuel might even be an invention, it is not an innovation until that solution can be practically deployed.

Innovation is the development of a new product or process in response to a defined problem or issue that can be practically implemented in such a way that it has a tangible benefit to the user, consumer, or societal entity that was originally identified as having the problem or issue.

ENTREPRENEURIAL MINDSET

When we use the term *entrepreneur*, what comes to mind? Is it something you see yourself being at some point in your career? If you already consider yourself to be an entrepreneur, how might you define the mindset of an entrepreneur? Why would being entrepreneurial be important when it comes to the design process and engineering design? Let's look at these questions a little more carefully, and see how we can define entrepreneurial mindset from within the framework of the design process.

Merriam-Webster defines the word *entrepreneur* as "one who organizes, manages, and assumes the risks of a business or enterprise."[10] Dictionary.com defines entrepreneur a little more broadly, stating "a person who organizes and manages *any* enterprise, especially a business, usually with considerable initiative and risk,"[11] with emphasis on the word *any* being our own. Both definitions might fit your interpretation of the word. For us, these definitions bring to mind the classical entrepreneurial story as told by the TV show *Shark Tank*, where person after person tells a story about their "a-ha" moment when a personal pain point was solved by an inspirational invention. They quit their job as vice president of a Fortune 100 company to make this dream a reality and have spent their entire savings attempting to bring their invention to market. This story ticks all the boxes of our dictionary definitions: managing an enterprise, showing initiative, and assuming all risks associated with that endeavor.

I think we can agree that many of these stories also talk about elements of the engineering design process that are almost always present: identifying the problem, brainstorming different solutions, prototyping, testing and validation, implementing the final design. So what do we mean by entrepreneurial mindset? What if you have no desire to be an entrepreneur?

Entrepreneurial mindset has little to do with how we previously described being an entrepreneur, at least in terms of actions; one might argue that the entrepreneurial spirit guided how we are going to define this mindset. This definition comes to us from KEEN (the Kern Entrepreneurial Engineering Network), which is a partnership of more than 50 colleges and universities across the United States that fosters collaborations among its network to implement entrepreneurially minded learning in a variety of contexts, and can be found prominently on their Engineering Unleashed website.[12] It is further from the Merriam-Webster definition than that of Dictionary.com, removing the entire concept of "enterprise" from the idea. Entrepreneurial mindset focuses on three C's: curiosity, connections, and creating value. By being curious about the world around you, making connections between things you already know and those that you are attempting to understand, and ultimately finding solutions that create value, be it within the framework of economic, societal, environmental, or even global impacts, this

---

10 "Entrepreneur," Merriam-Webster.com, accessed April 20, 2021, https://www.merriam-webster.com/dictionary/entrepreneur.

11 "Entrepreneur," Dictionary.com, accessed April 20, 2021, https://www.dictionary.com/browse/entrepreneur.

12 "Welcome to Engineering Unleashed," EngineeringUnleashed.com, accessed April 20, 2021, https://engineeringunleashed.com.

mindset can provide a way to link the engineering design process to innovation and design thinking. **Identifying opportunities is key for all great entrepreneurs, and if done correctly, the developmental phases of the design process can ensure the end result is an impactful design.** Again, this goes back to the premise we discussed earlier in this chapter that the key to a successful design is understanding what the real problem is.

An entrepreneurial mindset focuses on being *curious* about the world around you, making *connections* between things you already know and those you are attempting to understand, and developing solutions that *create value*.

What if you don't want to be an entrepreneur? How does this mindset help individuals that would prefer to work for a company or corporation? As a direct corollary, entrepreneurial principles applied within the context of a corporation are often defined as *intrapreneurship*. **An intrapreneur is someone who drives innovation from within the structure of a more traditional company.** Although being intrapreneurial may introduce some level of risk, that risk is taken on by the corporation rather than the individual, and typically these risks are managed in such a way as to not bring significant financial harm to the parent company should the innovation fail. In many cases, intrapreneurial entities are run independently or as part of a smaller subsidiary or offshoot of the parent company. Examples of this form of structure can be traced back to as early as 1943, when Lockheed Martin launched its Advanced Development Program (commonly referred to as "Skunk Works"), and continues today with companies using incubators or secret projects to physically and organizationally separate the innovation team from the day-to-day operations of the company. Even if you are not part of a formal innovation team, companies can directly benefit from you acting in an intrapreneurial way using the three C's we described when defining entrepreneurial mindset. **In creating value above and beyond your expected contributions to your company as described by your job description, independent intrapreneurship is a way to increase your own worth to your company.**

CHAPTER SUMMARY

This chapter focuses on the engineering design process, including a comparison of this process to both the scientific method and design thinking and the links between design and a broader definition of entrepreneurship. Within this chapter, the following topics were covered:

- Descriptions of each of the steps of the engineering design process
- The importance of recognizing, and attempting to minimize, bias in observations

- The need to establish criteria and constraints, gather data from reliable sources, and justify assumptions utilized during the process
- The concept of design thinking and the correlation of this to the engineering design process for "wicked" problems
- Entrepreneurial concepts that include the three C's of curiosity, connections, and creating value, and their relationship to effective engineering design

END OF CHAPTER ACTIVITIES

Individual Activities

1. Think back on a science experiment that you conducted at some point in your life. What question did you ask? Was the goal to understand a phenomenon or to solve a problem? If it was to understand a phenomenon, how could the experiment be adjusted to solve a problem? If it was to solve a problem, how might changing your test improve your results?
2. Select a spot in public where you can observe a large number of individuals in numerous and various situations. Select one of these situations and conduct observations, taking into account the six items presented in the observation section of the text.
3. List the top 3–5 regulatory agencies with which you will likely interact as an engineer. How might each of these agencies influence potential design solutions?
4. Consider a recent problem you have noticed or encountered on campus. Take ten minutes and brainwrite as many solutions as possible to the problem. The goal here is to consistently generate potential solutions for as much of the ten minutes as possible.
5. Watch the KEEN video on entrepreneurial mindset (https://youtu.be/WZHvRpuemgk). How do you believe you will use both your skillset and mindset to achieve your engineering goals?

Team Activities

1. The recent pandemic presented several challenges regarding vaccine delivery. As a group, create a list of criteria and constraints associated with vaccinations in the following locations. How might the same criteria or constraints present as different challenges in the different locations? Which criteria or constraints are issues in some locations but not others?
    a. New York City, New York, United States
    b. Coldwater, Kansas, United States
    c. Moore Haven, Florida, United States
    d. Paris, France
    e. Kampala, Uganda
    f. Manaus, Brazil
2. Conduct a search for the UN Sustainable Development Goals. As a team, select one of these goals and investigate further some of the targets, events, and actions associated with that

goal. As a team, create a single-page flyer that summarizes the current state of the goal, what has been accomplished, and what is left to be achieved.
3. Find a market segment you and your team are familiar with (or one that aligns with your chosen engineering major) that has multiple companies and where technology and innovation play an important role in advertising and market share (examples that immediately come to mind are a variety of consumer electronic goods, such as cell phones, computers, or televisions). Based on the market you've chosen, create a list of design inputs that define what that product must do for it to be a competitive product. Additionally provide weights to those inputs based on their importance. Research 4–5 products in that segment and rank their performance in satisfying the design inputs. Based on your research, which device would you purchase? Did you find any gaps in meeting the needs you identified, suggesting there might be an opportunity for further innovation?
4. Take three minutes to individually brainstorm a list of at least five things about which you are curious. Next, come together as a team to discuss the items on each individual's list and see if there are connections between items. What opportunities might exist to create value around items with the most connections?

CHAPTER REFERENCES

Barry, Keith. "The Crash Test Bias: How Male-Focused Testing Puts Female Drivers at Risk." *Consumer Reports*, October 23, 2019. https://www.consumerreports.org/car-safety/crash-test-bias-how-male-focused-testing-puts-female-drivers-at-risk.

Clancy, Tom. 1995. *Debt of Honor*. New York: Penguin.

Dictionary.com. "Entrepreneur." Accessed April 20, 2021. https://www.dictionary.com/browse/entrepreneur.

EngineeringUnleashed.com. "Welcome to Engineering Unleashed." Accessed April 20, 2021. https://engineeringunleashed.com.

Finley, Robert E. 1966. *Manufacturing Man and His Job*. American Management Association, Inc.

Merriam-Webster.com. "Entrepreneur." Accessed April 20, 2021. https://www.merriam-webster.com/dictionary/entrepreneur.

Ohno, T. *Toyota Production System: Beyond Large-Scale Production*. CRC Press, 1988.

Osborn, Alex F. 1953. *Applied Imagination: Principles and Procedures of Creative Thinking*. New York: Scribner.

Vlaskovits, Patrick. "Henry Ford, Innovation, and That 'Faster Horse' Quote." August 29, 2011. https://hbr.org/2011/08/henry-ford-never-said-the-fast.

CHAPTER 2
ENGINEERING SUCCESS

BEFORE YOU BEGIN

Take a minute to consider the following questions:

- What do you consider to be the critical components of a successful college career?
- Of the skills that you have developed in high school, which will contribute to your success in college?
- What are your main concerns regarding the current academic year?
- What are the characteristics of an effective team and an effective leader?
- What is academic integrity, and what are the most common ways in which it is not met?
- Who are some current members of your professional network?

LEARNING OBJECTIVES

By the end of this chapter, you should be able to:

- Recognize how the way you think and what you think impacts your achievements
- Create SMART goals and align these goals with long-term objectives
- Establish clear organizational and time management skills
- Prioritize various activities and correlate these to personal goals
- Identify academic integrity and potential intentional and unintentional violations to your university's academic integrity policy
- Summarize the characteristics of effective leadership and teamwork
- Develop an initial network of individuals and identify opportunities for adding to this network

INTRODUCTION

The concept of success can have a very different meaning for multiple individuals, and even a single individual can view the same result as success in one instance and something less than success in a different case. Consider someone training to run a marathon: the first time they run a six-minute mile may feel like a success but may feel much less so after training another five months. On the other hand, someone who is just beginning a running routine may consider themselves successful if they maintain a ten-minute mile—a threshold the marathon runner is unlikely to view as a benchmark of personal success. These differences in the definition of success go back to the particular goals that the two individuals have regarding running. The focus of this chapter is engineering success during your academic career, and it ties to the next chapter (Chapter 3), which focuses on continuing that success after graduation. Just as success is defined somewhat differently for the two runners discussed previously, we recognize that success and personal goals are different for every individual pursuing a degree. Because of this, we start this chapter with the "big picture" (or 14,000-foot view) of success before focusing on the 5,000-foot view and then the 25-foot view, with these closer foci providing guidance without losing sight (hopefully) of the larger picture.

METACOGNITION AND GROWTH MINDSET

Metacognition is the process of thinking about thinking, or understanding how we collect, store, and process information. Research[1,2] suggests that individuals who are aware of their thought process as they go about solving problems or learning new material are more likely to be successful in their activities. Simply being aware of how you should approach the concepts you are learning and continuing that in tandem with what you are learning makes the learning itself more effective. Along with this expanded perception of the thought process is also the critical nature of the manner in which an individual approaches learning.

Henry Ford once said, "Whether you think you can or you think you can't—you're right." This quote is, in some ways, a description of Carol Dweck's mindsets. Dweck[3] suggests that individuals fall into one of two categories: those with a growth mindset or those with a fixed mindset. Individuals with fixed mindsets believe that their skills and abilities are set (fixed) based at a prescribed level, and no amount of time

---

1 National Research Council, *How People Learn: Brain, Mind, Experience, and School: Expanded Edition* (Washington, DC: The National Academies Press, 2000), https://doi.org/10.17226/9853.

2 Ambrose, S. et al.. *How Learning Works: Seven Research-Based Principles for Smart Teaching*, ISBN 9780470484104.

3 *Mindset: The New Psychology of Success*, (2007), ISBN 9780345472328.

or effort can change that. In contrast, those with a growth mindset recognize that skills and abilities can be developed with time, practice, patience, and effort. **Growth mindset individuals consider failure a step in the learning process and welcome opportunities to challenge themselves.** Not achieving something one day simply means it needs to be tried again, maybe in a different way or at a later date. Studies have shown that individuals with a growth mindset (those who think they can) have demonstrably more success than those with a fixed mindset (those who think they cannot), particularly when faced with failure. Failure, though hard to accept, might be one of the key components to success. Those individuals who can look beyond the short-term negatives associated with failure and see instead the long-term potential for future gains demonstrate the growth mindset. This is not to say that individuals won't reach different levels of achievements (we all can't be Sheldon Cooper from *The Big Bang Theory* and *Young Sheldon*), but everyone has the ability to improve — "small, positive deltas," as one of our colleagues would say.

Those with a growth mindset recognize that skills and abilities can be developed with time, practice, patience, and effort.

PLAN FOR SUCCESS/SETTING GOALS

Now that we recognize the impact our thoughts (and how we think about our thoughts) have on reaching our goals, it is probably a good time to begin to think about exactly what our goals are. What is it that you hope to accomplish in this lifetime? When you look back fifty years from now, what would you consider a successful life? How does your degree and subsequent career play into that success?

Some of these questions may appear at this point in your life to be abstract at best. Certainly, they are things to strive toward, but how do you go about planning for fifty years from now when you are not even sure what you want to eat for lunch today? Perhaps you are not even sure if engineering is the right career path for you at this point in time. Beyond these uncertainties, which you might have some level of control over, given how quickly the world is changing, how do you account for things that appear to be indirectly or not at all under your control?

Our suggestion regarding this is to develop a range of short- and mid-term goals that reflect your longer-term aspirations. The framework we will be presenting in this section can easily be adapted to your specific needs, and as you will see it is developed in a way that allows and expects frequent evaluation and adaptation. In order to fully understand what purpose this framework will have in helping you reach your aspirations, we first need to introduce you to the concept of SMART (**s**pecific, **m**easurable, **a**chievable, **r**elevant and **t**ime-based) goals. SMART goals have five key attributes to them, that ultimately create the mnemonic or acronym SMART.

SMART goals are specific, measurable, achievable, relevant, and time-based.

1. **Specific.** Narrowing down the focus of your goal provides you a better understanding of how the goal will be accomplished. Instead of saying "I want to do better in my classes," try to be more specific about the term "better," perhaps by aspiring to a specific grade holistically within the class or even looking at your grades within a specific deliverable. "I want to get a B or higher in all my classes," or, "I want to get all A's and B's on my exams," provides a clearer definition of what you mean by "better." As we will see by our next point on this list, this type of specificity also allows our goal to be measurable.

2. **Measurable.** For a goal to be SMART, you need to be able to measure it. Looking back at our example of "I want to do better in my classes," by not defining the term "better," you don't allow yourself to measure your success. If you got a 70% on your first chemistry lab, is "better" a 71% on the next lab? Your score improved, but did that meet your expectations? Without allowing for a measurable outcome, you may be uncertain if your goal was achieved, or it may provide you wiggle room to declare success even if that success was not sufficient to attain your desired outcomes. In the example of an improvement of 1% in your chemistry class, if your aspiration or desired outcome was to earn a "B" in the course, neither overall score likely helped in your attainment, even though you did "better."

3. **Achievable.** While you want to make sure any goal you set challenges you to do your best, you also want to make sure that those goals are reasonable in regard to their scope and expectations. This rationale should be based not only on your capabilities but also other constraints your SMART goal may require (e.g., how much time you have to achieve your goal). For example, if your college organization fundraiser raised $20,000 during its last fundraiser, and you decided that in order to challenge yourselves, you set a goal of $250,000, the goal might not be achievable given the evidence of your previous efforts. If efforts for the last several fundraisers all hovered around $20,000, perhaps an appropriate, yet challenging goal would be to increase your fundraising efforts by 50% for the upcoming event, setting your goal at a more realistic $30,000.

4. **Relevant.** For many, this next element of a SMART goal goes without saying, but obviously any goal you set should be relevant to your objective or outcome. Obviously if your desired outcome is to make the dean's list, having a goal to tutor students struggling in physics, while noble, may not be relevant. While this example is reasonably black and white, the relevancy of some goals may be much more subtle. Regardless, make sure the goal appropriately feeds back to the desired outcome.

5. **Time-Based.** Of the five attributes, understanding the time frame in which the goal will be active is perhaps the most important aspect of SMART goals. This attribute plays an important part in determining what results are achievable and provides a timeline for measurement and reflection. As we will see below, a good plan will likely have multiple goals, each of which may have different timelines for completion and assessment. This type of plan provides an opportunity for continual assessment and feedback and allows for goals to stretch for longer periods of time.

Putting each of these pieces together and developing a truly SMART goal takes time and practice. Keeping the five key attributes in mind, one can take a goal that is difficult to measure and assess, "I want to do better in my classes," and develop something that is much more meaningful and impactful in assessing your progress. "I want to get at least an 85% on all exams in my calculus class this semester." Once you have a SMART goal in place, you will ultimately be able to develop an action plan that will provide you a path as to how you will achieve that goal. In the case of your calculus exams, you might state: I will spend at least one hour looking over any pre-class materials and readings prior to each class and at least one hour after each class going over course notes and sample problems. In addition, I will spend at least three hours each week working through homework problems and attending office hours and recitations as necessary. One week prior to each exam, I will spend two hours per day reviewing all necessary materials for the course.

> **Engineering in Action**
>
> Consider one of your main concerns for this academic year that you thought about in the "before you begin" questions. Develop a SMART goal that can potentially reduce this concern. Once you have created your goal, pair up with a classmate and present your goals to each other. Can both goals be considered SMART goals? How might they be changed if they do not meet all five SMART characteristics?

Now that you have some practice creating SMART goals, we want to use these as inputs into your success plan. At our own home institution, we have our first-year students complete a *student enhancement plan*, or SEP, which focuses on activities in three specific areas we find pertinent to most students: academics, professional development, and civic engagement. We recommend students revise or rewrite their SEP each year so they have a roadmap of what they need to accomplish in the upcoming academic year to meet their academic and career aspirations. A template for this document is provided in Figure 2.1.

As you can see from Figure 2.1, the SEP begins by including a strategic objective which will ultimately define the rest of your document. We recommend focusing on mid-range (i.e., three to four years) and long-range (five or more years) objectives, which would typically correspond to graduation (mid-range) and what you currently envision doing in your career several years post-graduation (five or more years). As with any goal-oriented document that spans multiple years, there will be a fine balance between flexibility (plans, objectives, and goals change with time) and accountability (goals should not change as a direct result of not meeting a specific expectation). Having a mix of short-, mid-, and long-term goals and objectives inside this "living, stretch"

> ## *Your Name*
>
> *{Major, anticipated graduation date}*
>
> *{College or University}*
>
> Student Enhancement Plan (SEP) for *{academic year}*
>
> ### Strategic Objective
> This section should provide a brief statement of your projected short- and long-term goals. Short term goals look at what you want to do upon graduation. Long term goals should reflect thoughts on where you would like to be in 5 years or so.
>
> ### Anticipated Roadblocks
> This section should provide a brief statement of what you think might keep you from meeting your strategic objective. This section should only be a paragraph or two.
>
> ### Academic Goals
> Goals in this section should focus on academics.
>
> ### Professional Development Goals
> This section should focus on professional development activities geared towards your short-term goals presented above.
>
> ### Civic Engagement Activities
> The can also be considered as service, but civic engagement implies a more intentional involvement. Service is wonderful, but time is valuable. Make sure activities are meaningful.

FIGURE 2.1 Template for a Student Enhancement Plan (SEP)

document —"living" in the sense that edits and updates will be necessary as the academic year progresses, and "stretch" in the sense that certain goals may span across multiple years—should help in achieving this balance. The document also includes a brief section where you are able to reflect on what challenges might prevent you from achieving your goals. This can be items such as having to take on additional hours at work, an increase in family commitments, or potentially a health issue with which you are contending. While not designed to focus on the negative, this section recognizes that at times things outside of your control can shift the trajectory of your path.

Once you have defined your strategic objective, you now need to develop a series of SMART goals in each of the three areas previously mentioned (academics, professional development, and civic engagement) that will help set a course for your upcoming academic year. **These goals will ultimately be your roadmap to success, and as such careful thought should be put into**

their development. We encourage students to try to think of this document as more than just a checklist of things to do; in fact, we highly recommend developing goals that range across the spectrum of achievability (i.e., from nearly certain to achieve to aspirational in nature). Having this mix will encourage you to continue to push yourself, recognizing that even if all your goals are not met, you are hopefully closer to achieving your objective at the end of the academic year than you were at the beginning of the year. Some goals may span across multiple years and will help in maintaining continuity across each year's SEP. Numerous options exist for these goals, and there are no right or wrong options for topics to include. Remember that this document is for you. You ultimately define your success in achievement, and what you get out of this type of planning and assessment is directly proportional to the amount of effort you invest and how honest you are with yourself in that assessment.

> **Engineering in Action**
>
> As a group, brainstorm examples of goals that would fit into each of the SEP categories: academic, professional development, and civic engagement. Consider short-term goals that may be appropriate currently, as well as ones that you might consider later in your academic career.

ORGANIZATION AND TIME MANAGEMENT

The SEP that you develop is a great guide for compiling your goals but may or may not assist in the day-to-day accomplishment of those goals. We mentioned in the last section that it is a good idea to check back periodically with your SMART plan to see how you are progressing toward your set goals. If you find you are doing well during these checks, that is excellent; if you find instead that you have not made much progress or you are not satisfied with where you are, it may be a good time to consider your personal levels of *organization* and *time management*. We tend to think of these two items together, but that doesn't necessarily mean they are directly related. It is possible to be highly organized but not necessarily have great time management skills. We do believe that being better organized makes it easier to manage your time, and time management plays into your overall personal organization. Let us take a minute to talk about each of these independently.

    Organization is the ability to identify and locate the many things we have going on. This is more than remembering where we put our keys or establishing a habit of making sure our phone is charged; it is also about recognizing what needs to be completed, when things are due, where we are supposed to be at a given time, and how many things we have going on. Whether it is extensive to-do lists or constant reminders on their phones, organized individuals are "in the know" when it comes to the things they need to accomplish. Being able to complete that extensive list, however, lies more with time management.

Time management is how we fit all our "to-dos" into the time we have. On a personal level the authors often talk about how much more we could accomplish in a 26-hour day, but as yet we have not found a way to create those additional two hours. While we tend to say this to each other in jest, particularly when our schedules get inordinately busy, it speaks to the reality that we all face: there are only 24 hours in a day. In order to effectively accomplish everything, we need to take a very honest view of how we spend those hours.

Try the following activity:

1. Make a list of everything you would like to accomplish each typical weekday.
    a. Be sure to include everything (sleeping, eating, working, studying, class, watching Tik Tok, working out, etc.).
    b. If you have specific, regular weekly activities, include these in the list as well.
    c. Divide a sheet of paper into three columns. Place the items in the center column, each on a separate line.
2. To the left of each item, rank each item in terms of importance.
    a. Use the scale of very high, high, medium, low, or not important. There is no right or wrong here; this is what is important to you.
    b. Once you are done, **fold over the left side of the paper so that these rankings are no longer visible**, but your list of items can still be seen.
3. To the right of this list, include the amount of time you typically spend on each activity.
    a. **Focus on the time you actually spend**, not the time you feel you should spend or want to spend.
    b. Round to thirty-minute chunks. We recognize this is not completely precise, but it will give you a general idea.
4. On a second sheet of paper, start with 24 hours at the top. Look at your initial list and select the item you do most often. Note that **this may not be the one that takes the most time, but rather the one that you are guaranteed to do every day**.
    a. List this item first, along with the time you estimated you spend on it.
    b. Subtract that time from 24 hours and note how much time you have left.
    c. Continue to repeat this process. The first few items on this list will probably fall into a "needs" category (eating, sleeping, etc.). If you are struggling to consider what to include as you get further down the list, think back on the last two or three days and include items that you did every day, followed by ones you did most days.
    d. Make sure to subtract the time each time you add something to the list. For example, if the first item on your list is sleep (7.5 hours), you would have 24 − 7.5 = 16.5 hours. The second may be meals (2 hours), which then takes you to 16.5 − 2 = 14.5 hours.
    e. Add items to the list until your time drops to zero or as close to zero as possible without going negative.

5. Once your second list is complete, place both lists side by side.
    a. Unfold the left side of your first list so that your importance rankings are now visible.
    b. Compare your two lists. How many of your very high and high items from the importance list made your daily list? How much of your time was spent on low or not important items? Which (if any) of your very high importance items did not make the daily list?

A second activity you can try is to take the same list that you developed in steps 1–3 and create a new daily (24-hour) sheet. This time, select the items that you deemed "very high" in importance, list these, and perform the subtraction process. Once you have included all of the very high items, move to high items and down the importance list until you run out of hours in your day. How far down your list did you get? Were you able to include all your very high and high importance items? What about those not important items? Out of curiosity, if you added up all the items on your initial list, how much time would they take?

> **Engineering in Action**
>
> In pairs or as a small group, discuss with each other what you discovered with the previous activity. What did you learn? Were you surprised by your results? How well did your actual times and your target times match up? Did you discover any "time sinks"? Do others in your group include items that you forgot to consider? Are there commonalities across your experiences?

Activities such as these can begin to highlight potential differences between what we consider to be most important and what is most time-consuming in our lives. While many things may fall into both the important and time-consuming categories, almost all of us have activities that are time-consuming but that we do not consider to be very important. Some of these items may be things we do not have much control over. For example, our commute may take us forty-five minutes each way. We may not consider these items terribly important, but we also may not be able to make changes to them without significant consequences. Other items we may have notably more control over. Maybe we enjoy sleeping late but can reserve that for weekends or can choose to limit our Hulu watching to two episodes rather than five of our favorite show. This is not to say we should never relax or that everything we do needs to be very important. But if we find ourselves struggling to reach our goals, **taking an honest look at how we spend our time can potentially provide opportunities to realign our activities with our priorities**. Many student success centers have resources for students on developing and maintaining time management skills. We encourage you to check out yours or simply Google "time management tips."

CLASSES

In the previous section, you were asked to consider what activities were the most important and the most time-consuming in your life. Assuming that you are reading this as a student, classes were probably on the upper portion of one or both of those lists. Whether you are full-time or part-time, in person, online, or a combination of the two, **classes and the associated work plays a major role in your academic career**. Most colleges and universities have established a specific curriculum (or set of classes) that you are required to take to earn your degree. These classes were established with the creation of the program and are influenced by state standards, college or university expectations, regional accrediting bodies, and specific program accreditations. Depending on the specificity of each of these groups and the size and age of your program, the courses you must take for graduation may be very specific or a bit more flexible. Classes tend to fall into one of three categories:

- **Required** courses are those that are, as the name implies, required for your degree. These are set as a specific course, and thus substitutions are rarely allowed. Examples of required courses in an engineering program are often Calculus I, Chemistry, Statics, Fluid Mechanics, and Senior Design/Capstone to name a few.
- **Restricted Elective** courses are courses that have specific restrictions. These can fall into broad categories (for example, a general education course that is classified as a social science) or be a choice from a set list of courses (e.g., pick 1 of 3 or 2 of 5 from the following list). Most general education courses fit into this category, depending on how your school approaches the general education requirements. Expectations of a set number of credits for humanities, social science, foreign language, writing-intensive, and so on are part of this group. In addition, many engineering programs have restricted electives in the upper level as well. Some have tracks or some other means of providing a recommended set of courses to take if you would like to specialize in a particular sub-discipline. Others simply note a requirement for a set number of credits of courses with a certain prefix, at a certain level, or from a generated list. Occasionally, these may be as specific as an either/or option. In any case, the purpose is to ensure that certain criteria are met while allowing some flexibility in topic coverage.
- **Free Elective** is a completely open course. There are no restrictions on which courses can count for this category, although the classes themselves may have restrictions (for example, a scuba diving course may require you to pass a swim test to enroll). Depending on your high school preparation, these free elective hours may be filled by courses needed to prepare you for required courses (e.g., Precalculus is often not included in an engineering curriculum, but if you did not have the opportunity to take it in high school, you would need to register for this course before taking Calculus I, so it would then serve as a free elective course). If these free elective hours are not met with pre-requisite courses not included

in your curriculum, they then become open opportunities to pursue topics of personal interest. Your school may have recommendations for courses to take to fill these credits or may leave it to you to investigate options. Some students choose to take advantage of this opportunity by augmenting their curriculum with additional courses that are aligned with their degree, while others take a different approach and enroll in a course for personal interest (e.g., photography, philosophy, music). We encourage you to take a course based on the topic—pick something you believe you will enjoy and are interested in learning more about—rather than based on the time the course is offered or the impact it might have on your GPA. An interviewer a few years down the road may be looking over your transcript and ask, "So why did you take …?"; you do not want to have your answer be, "It was an easy A that let me sleep late."

In addition to the type of class (required, restricted elective, free elective) a course is, other important course characteristics are when it is offered and its prerequisites. By *when* we do not mean 8 a.m. or noon (although we recognize that may be important to you as well) but rather in which quarter or semester the course is offered. Some classes are offered every session—fall, spring, summer—and may even have multiple sections in each session. Others are offered only once a year, and some may even be offered on alternating years or only occasionally. The frequency of the course offering and the number of seats in the course influences how easy or difficult it may be for you to register for the course when you want to take it. It is also important to make sure that you have the prerequisites for the course prior to the session in which you want to register. This leads to our recommendation to **plan out your course schedule early for your entire college career**. Obviously, you will not be able to select specific days, times, or instructors for every quarter or semester, but there are several benefits to having a general plan of courses for your degree.

First, this gives you a clear idea of a graduation date. While most undergraduate degrees are designed to be able to be completed in four years, there are several factors that can influence that timing. Going part time rather than full time; alternating school and work if you are in a co-op program (more on this in Chapter 3); having to take prerequisite courses prior to enrolling in your specific program; taking off a semester to work, travel, or study abroad; or having to drop classes due to health issues or injury are some of the events that can change this timing. Some of these can be predicted, some are intentional, and some are accidental or unexpected. Planning your complete course schedule cannot address the unexpected but can provide a better picture for those known factors. One of the best examples is a student who has outside obligations (work, family, etc.) that limit full-time attendance. At first glance, one may think only needing two or three classes rather than four or five would be easier to schedule rather than more difficult. But with prerequisite requirements and select courses not being offered every session, it is even more important to schedule further out for these situations. Taking it session-by-session may work for a year or so, but without a clear plan, you are much more likely to end up with a semester where potential courses cannot be taken (e.g., not offered or the prerequisites have not been completed) than if you were more intentional in your planning. The added benefit of this pre-planning is that concept clusters (i.e., courses that feed logically into one another) can be grouped together rather than spread out with large gaps between offerings. Nothing is more frustrating than acing your statics course, only to take mechanics of materials a year or more later and recognize that you have to go back and brush up on the material from statics that once was completely clear in your mind. While some refresher activities might be needed with courses spaced out more, there are options that minimize this frustration by clustering courses of similar topics into specific semesters. Early planning of classes for your entire career ties back to strong organization and clustering courses (if necessary) relates directly to time management.

ACADEMIC INTEGRITY

As mentioned in the Introduction chapter, being an engineer is about making the world a better place to live. Within any engineering society or organization, their oath or motto will use the words *integrity, standards,* and/or *honor*. Being an engineer is a great responsibility, and part of the expectation of being a practicing engineer is that you will act with integrity in all activities you pursue, whether on or off the job. While the National Society of Professional Engineers does not support automatic denials of licensure for applicants with criminal records, they do recognize that "there may be certain instances where criminal act(s) bear on the ability of the individual to engage in the practice of engineering and therefore should be weighed as part of the state engineering licensure process."[4] As an engineering student, these expectations carry over into your academic career as well.

Every institution has some sort of academic integrity or honesty policy associated with their rules for student conduct. For some, you may have participated in signing an honor pledge, either when you started at your university, or perhaps as part of the process for a specific class or exam. For others, perhaps your institution does not have an honor code per se, but instead defines the expectations regarding academic integrity as well as the potential ramifications should those expectations not be met. At our institution, the academic integrity policy is part of a larger student code of conduct, which covers a variety of conduct expectations that go beyond the material we will be covering in this section.

While it should go without saying that being academically honest in the pursuit of your education should be expected of every student, as professors, we understand there will be times during your academic career when acting outside or at least on the periphery of these ideals will seem tempting. A 2012 survey showed that 33% of high school students copied an internet

---

[4] https://www.nspe.org/resources/issues-and-advocacy/professional-policies-and-position-statements/engineering-licensure.

document at least once for a homework assignment, 52% of students cheated on an exam, and 74% copied someone else's homework.[5] At the college level, 21% of surveyed undergraduate students admitted to cheating on a test, while 62% admitted to "cut and paste" plagiarism on a written assignment.[6] It should be noted that these statistics were from a survey administered from 2002–2005, which, in the age of the internet and devices delivering instant information, is akin to the Stone Age. To some reading this, these numbers may seem higher than expected, to others perhaps too low. Regardless, these data suggest that academic dishonesty is somewhat commonplace in higher education. Of course, the argument that "everyone is doing it, so perhaps I should too" does not work in education, just as it does not work in professional sports (Lance Armstrong[7] and the Houston Astros[8] are excellent examples of this) or in business (mobile processor manufacturer MediaTek was found to have been less than forthcoming in their chip benchmarks[9]). How you choose to respond in those times and whether your conscious tells you something is okay or not is something you will eventually need to live with. At this point in your career, what you choose to do or not to do ultimately has little impact on anyone other than yourself and perhaps anyone you are working with if the activity in question involves more than just yourself. **Recognize, though, that acting without integrity can have significant consequences as a practicing engineer.**

The statistics provided point to what people most commonly view as academic misconduct, specifically cheating on an exam or written assignment. These forms of cheating as described are pretty straightforward; there really isn't any ambiguity as to the intention of the cheater. That being said, there are more subtle forms of academic misconduct that may not necessarily immediately come to mind when one talks about "cheating" or "academic misconduct." Actions such as free riding (working on a team project but failing to participate in the team activities equitably) and submitting your own work to two or more distinct classes (self-plagiarism) are forms of academic dishonesty that may not immediately come to mind when you think about cheating; nonetheless, they are activities you should be thoughtful of when discussing this topic.

---

5 https://www.drake.edu/raycenter/projects/education/k-12charactereducation/reportcardontheethicsofamericasyouth/.

6 Donald L. McCabe, "Cheating among College and University Students: A North American Perspective," *International Journal for Educational Integrity* 1, no. 1 (2005).

7 https://www.forbes.com/sites/hbsworkingknowledge/2013/12/18/lessons-from-the-lance-armstrong-cheating-scandal/ - 71ca50316902.

8 https://www.newsweek.com/former-houston-astro-says-if-mlb-probes-sign-stealing-other-teams-theyre-going-find-lot-more-1486814.

9 https://www.forbes.com/sites/moorinsights/2020/04/14/mediatek-refuses-to-back-down-after-being-caught-cheating-on-benchmarks/ - 7f446c107742.

> **Engineering in Action**
>
> Look up your college or university's academic conduct policy. This can often be found online, and many times a reference will be included in your syllabus. What is listed or described in that document as activities that violate the academic integrity policy? Did any of these activities surprise you? What are some of the potential consequences for violating the academic integrity policy?

So far, we have talked about academic dishonesty or misconduct, but the title for this section of the book is academic integrity. Are these terms simply the opposite of one another, or is integrity more than appropriate conduct or academic honesty (i.e., the opposite of dishonesty and misconduct)? If you look at how academic integrity is defined, per Webster's dictionary, it is a firm adherence to a code of especially moral values.[10] By definition, this goes well beyond appropriate academic conduct and focuses on the entirety (in terms of academic integrity) of the academic institution. As presented by the International Center for Academic Integrity (ICAI), academic integrity is a commitment to five fundamental values: honesty, trust, fairness, respect, and responsibility. Furthermore, teachers, learners, and researchers must demonstrate the courage to act in preserving these values in the face of adversity.[11] Go back to the creeds, oaths, and charters of the various engineering societies and professional organizations you examined in the introduction of this book, or even look at a company's code of conduct for their employees. These ideas and words will be found throughout all those documents. Even simplified versions of how to act with integrity, such as Google's famous "Don't Be Evil," focus on the holistic concepts described by the ICAI. Adhering to these values and not wavering in the face of adversity is something everyone should strive to meet throughout their daily lives, both inside and outside the classroom.

LEADERSHIP/TEAMWORK

One thing students quickly realize when they first get into an engineering course is how team-oriented the courses and discipline really are. There are many stereotypes of engineers being the awkward, loner sitting in his/her/their cubicle working on calculations no one else in the office can understand (think of Howard from *The Big Bang Theory* or the cartoon character Dilbert). This can come as a surprise to some students and can lead to additional stress as it requires them to rely on classmates to complete tasks within a course. In this section, we will provide

---

[10] https://www.merriam-webster.com/dictionary/integrity.
[11] T. Fishman, "The Fundamental Values of Academic Integrity," International Center for Academic Integrity.

some insight we have gained in regard to successful teamwork and how leadership is not only a characteristic employers look for in a potential job applicant (more on that in Chapter 3) but is also a necessity for **all** members of a team if a collaboration is to be successful.

If your first reaction to hearing your instructor say, "Get into your groups and work," is general apathy or perhaps even a level of resentment or displeasure, you are not alone. In our experience, group work is one of the things students most often complain about, and the reasons for their complaints are many. As faculty, we recognize that while group work should result in a product that is greater than the sum of the individual efforts, the actual result is usually much less than this and perhaps even less than the sum of the individual efforts. For a group to work well, it must overcome a number of obstacles, including appropriate coordination and delegation of work, ensuring all team members are appropriately motivated to complete the task(s) at hand, and ensuring that the group remains productive over time.[12] Each of these issues are hurdles that groups will have to overcome in order to be successful, and rather than trying to provide you with some generic strategies that may or may not work (unfortunately, there is no one-size-fits-all solution), we thought it might be best to appropriately define these obstacles so that you can be aware of what causes them and perhaps figure out ways within your own groups to mitigate potential pain points early in the process.

Coordination and Delegation of Tasks

The first hurdle your team will likely face is the five W's of team meetings: who, what, where, when and why. Determining how your meetings will take place (in person or online, synchronously or asynchronously, etc.); coordinating schedules, particularly if your group is large; determining how decisions will be made, including how work will be divided among the group; and how individual contributions will be integrated into group submissions are all things that will need to be considered upfront and worked out in order for your group to be successful. Typically, when groups are unable to overcome this hurdle, tasks end up incomplete, deliverable deadlines are missed, and the work that is completed is hastily put together and not well integrated. This is one area where strong leadership can develop a plan of action that includes delegated responsibilities amongst teammates and accountability measures for those who do not meet agreed-upon deadlines or work quality standards. Generally speaking, if your group cannot get past these early roadblocks, the opportunity for success is greatly diminished.

Motivation

Motivation is the issue we probably face most often in our own classes, and it is one that is very dependent on group dynamics and the personalities of the group members. In our previous section on academic integrity, we describe the concept of free riding, which is allowing others to complete most or all of the work while you (if you are the free rider) do very little or

---

[12] https://www.cmu.edu/teaching/designteach/teach/instructionalstrategies/groupprojects/challenges.html.

nothing. While this phenomenon, as written, is describing a single individual not pulling his, her, or their own weight, there is a similar situation where the entire team begins to lack motivation, assuming that someone else in the group will complete the task. This can lead to an overall decrease in the amount of work completed and the level of productivity the group demonstrates on a project. The third phenomenon we see occur regularly associated with group motivation is the inability to resolve conflicts. These conflicts might result from actual or perceived free riding but may also be due to other situations, sometimes spilling over into relationships within other courses or even people's personal lives. As we suggested for coordination and delegation, having strong leadership within the group can mitigate some of these issues. Ensuring that responsibilities are delegated appropriately, having deadlines and quality standards in place, and having accountability measures for those who do not meet the standards agreed upon by the group are common methods used to prevent these issues from occurring. Having experience in or knowledge of skills associated with conflict resolution is valuable and may be taught as part of your coursework in certain classes. Additionally, while your instructor might use peer evaluations to assess teaming skills and motivation as part of the external evaluation of your group, similar methods of assessment, including self-evaluations, can be used by your group independent of the external assessment to gauge motivation among team members.

Having strong group leadership can often mitigate issues of coordination and motivation in group settings.

Productivity

While coordination and delegation of tasks often occur early within the timeline of teamwork, the productivity issues we are going to describe here are phenomena that typically occur much later in the timeline (motivation has the ability to appear at any time), and often occur in groups that actually may be working well with one another. One such phenomenon is when a group becomes so set in its ways, as is particularly true for longer projects, that it becomes detrimental; they continue to move forward with plans and strategies even when they are not working and are unable to pivot or reexamine the situation. There are multiple examples throughout recent history where companies were unable to adapt to changing technologies or consumer sentiment: Kodak upon the development of digital photography[13]; Blockbuster, which failed to recognize the market's desire to rent DVDs via delivery (Netflix was a DVD rental company before streaming became popular)[14]; and Yahoo, which focused on the development of media and wasted away

---

13   https://brand-minds.medium.com/why-did-kodak-fail-and-what-can-you-learn-from-its-failure-70b92793493c/.

14   https://www.forbes.com/sites/gregsatell/2014/09/05/a-look-back-at-why-blockbuster-really-failed-and-why-it-didnt-have-to/#1c2d73d11d64/.

their advertising market.[15] Similar to those examples, groups that fail to take stock of their current work and look for ways to improve efforts or ideas typically end up with a less than ideal final product, as they allow inefficiencies or even deficiencies to continue unabated.

Another way group dynamics can negatively impact productivity is by displaying a groupthink mentality. Groupthink is a process by which a team or group of individuals makes decisions based on reaching a consensus, often resulting in suboptimal decision-making. In these situations, a consensus is so important to the individuals within the group that they are willing to sacrifice speaking up about something they disagree with. As a result, critical examination of decisions and exploration of other options are put aside for the good of the team. This "don't rock the boat" mentality can lead to poor decisions that, to an outside, objective observer, seem impossible to comprehend.

In both phenomena, leadership, communication, and organization can help avoid some of these roadblocks. Avoiding complacency, critically evaluating decisions and work, and searching for ways to continue to get better are all hallmarks of strong leadership, and they are also methods to prevent decreases in productivity. Although most group projects will likely have some form of a team leader, leadership qualities need to be exemplified in all team members in order for a team to have success. While trusting your teammates to follow through with their assigned tasks is not always easy, making sure each person on the team is held to a level of personal accountability is something you will likely be doing time and time again throughout your undergraduate career and beyond. Learning the skills necessary to be a valued teammate and contributor to a project is one of the hardest things engineering students are asked to do.

In order for a team to be truly successful, all team members should actively exercise effective leadership qualities, not just the single team leader.

CAMPUS ENGAGEMENT

While coursework plays a large part in the college experience, classes are not (and should not) be the sole focus of your time in college. Do not take this to mean that we are encouraging you to skip classes or blow off homework—success in your courses is vital to earning a degree—but recognize that there is value in activities, events, and engagement outside of the classroom. Campus engagement can take many forms, and the level of this engagement crosses the spectrum from absolutely none to what we would probably characterize as so-great-as-to-be-detrimental. The goal is to avoid the extremes, as neither of these will benefit you in the long term. Aside from that, where you fall on the spectrum is a completely personal decision. In this section we would

---

15 https://www.reuters.com/article/us-yahoo-m-a-missteps-analysis/the-identity-crisis-that-led-to-yahoos-demise-idUSKCN1060DN/.

like to make you aware of some of the opportunities that may exist on your campus, and things to consider as you look to be involved.

While coursework plays a large part in the college experience, recognize there is value in activities, events, and engagement outside of the classroom.

Campus engagement opportunities can be categorized in several different ways. We will split these into two main categories focused on major-related organizations and personal interest organizations. We are aware that this list is not all-inclusive. If we tried to address every opportunity on campus, it would probably double the length of this chapter and take time away from actually experiencing some of these opportunities, but this section will provide you with a solid starting point for your own exploration.

Major-Related Organizations

In the Introduction chapter, we mentioned how each branch of engineering has a related professional organization. These organizations are often at the national or international level, but they have state and regional chapters. Examples of these include the American Society of Mechanical Engineers (ASME), the American Society of Civil Engineers (ASCE), or the Institute of Electrical and Electronics Engineers (IEEE). In addition to the professional chapters, most also have associated student chapters that exist at colleges and universities educating students pursuing these various degrees. These student chapters have their own activities separate from the "parent" chapters (nearby professional chapters) as well as joint activities with the parent chapters. Most parent chapters also offer opportunities for individuals from the student chapter to participate in most of the parent chapter events. In addition to discipline-specific organizations, other major-related groups can include the Society of Women Engineers (SWE), the National Society of Black Engineers (NSBE), the Society of Hispanic Professional Engineers (SHPE), and Engineers without Borders (EWB), just to name a few.

> **Engineering in Action**
>
> Look on campus or on your college or university website for various engineering student organizations. Find one or two that look interesting and contact them or attend the next open meeting. You will likely find they are welcoming of new members and excited for you to get involved.

Personal Interest Organizations

While we hope that all of you actually have a personal interest in your major, by personal interest organization, we mean something independent from your major. Whether this is engaging with a Greek organization, joining the school quidditch team, or becoming a member of the improv group on campus, getting involved with a group of individuals outside of your major is valuable on multiple levels. Remember, college is an experience: meeting new people, learning new things, and gaining different experiences. It is not necessary to be focused on engineering twenty-four hours a day, seven days a week. Working and playing with people who have similar interests (even if different majors) can be relaxing and enjoyable and may even be educational in a completely different way than you expect. One of the more unexpected pairings we have encountered recently at our home institution is a collaboration between engineering students and art students whose goal is to repurpose concrete canoes into a campus art project in one of our many lakes. Becoming an engineer (and studying to be an engineer) does not mean you have to give up other interests. Take the time to stay engaged with that morning intramural basketball game, spend time at a local soup kitchen, or dust off your trumpet and try out for one of the jazz quartets.

Timing for Campus Engagement

We recognize that time plays a huge part in your level of campus engagement. Therefore, we suggest developing a student enhancement plan (SEP), which we introduced earlier in this chapter. Considering, at least in part, how activities contribute to short- and long-term goal achievement can help us to prioritize activities, events, and obligations in our lives. Any campus engagement must be balanced with the following in mind:

- sufficient time for class requirements
- any part- or full-time work requirements
- personal family obligations you may have
- time commitments associated with training/practice/games if you are a student athlete
- required attendance at military training, drills, deployment

Think back on the time management activity that was presented earlier in this chapter. Remember to compare your needs versus wants, and consider what may be a time sink versus those things that take time but result in big payoffs academically, socially, physically, spiritually, or mentally. **Remember that the way you commit your time may vary over the semester and can definitely change from semester to semester,** so look to re-evaluate and revisit your SEP at the start of each semester or year to make sure you are staying on target with your personal short- and long-term goals.

FACULTY/ADVISOR RELATIONSHIPS

As we have said repeatedly in this chapter, your college experience goes well beyond the courses you take. Your involvement in activities outside the classroom in many ways is as important to your future endeavors as is your pursuit of knowledge within the classroom. Relationships, not only among your peers but also with the faculty and advisors in your department, school, and college, can be extremely important in helping you define your career pathway. Faculty and advisors offer a wealth of knowledge that can provide you additional perspectives when it comes to making decisions regarding your professional development. Getting to know these individuals and letting them get to know you can lead to information and opportunities that otherwise may not have been apparent to you.

Connections with peers, faculty members, advisors, and mentors can help you define your career path, but you ultimately need to be an advocate for your own education.

As with any of the sections in this chapter, what you get out of faculty or advisor relationships is likely to be directly proportional to the amount of effort you put into the relationship. One big difference in the transition from high school to college is that you now need to self-advocate for your education and professional development. Opportunities within a college campus are nearly limitless; however, almost all of them will require you to seek out the engagement opportunity rather than expect the opportunity to come to you, asking for your participation. An example we see all the time as faculty is our own office hours. Most faculty set aside several hours a week in their schedules to allow students in their classes the opportunity to ask questions or get further clarification about their course. Unfortunately, surveys have shown that nearly two-thirds of students never participate in office hours,[16] and that students view these opportunities for faculty engagement as something that either should be reserved for emergencies or something that is not worth their time and effort because a similar outcome can be achieved via email. While these viewpoints may be valid if all you are interested in is getting the right answer on a homework problem or gaining a better understanding of the derivation that was completed in class, our argument here is that there exists an opportunity to transition this student–instructor relationship to a mentee–mentor interaction that can ultimately be more rewarding for all parties involved.

One comment we hear frequently when talking with students is that they feel their questions are trivial, and they do not want to "bother us" or "waste our time" answering questions the

---

[16] Margaret Smith, Yujie Chen, Rachel Berndtson, Kristen M Burson, and Whitney Griffin, "'Office Hours are Kind of Weird': Reclaiming a Resource to Foster Student-Faculty Interaction," *InSight: A Journal of Scholarly Teaching* 12 (2017): 14–29.

students feel they can ultimately figure out on their own. While it is true that faculty, like most folks, are busy individuals, if we set aside time in our schedule to be available to answer questions, we do so anticipating that students will use that time to ask questions. While we always have other things we could be doing during that time (just as you engage in other activities while waiting in line), our intent is to provide a means by which students can access us outside of the classroom, hopefully in an environment that is not intimidating to the student. Admittedly, some faculty may come across as being succinct (sometimes to the point of bluntness) in their answers either in class or via email; however, often this is in response to multiple questions or emails that need to be answered, and what may appear to you as being direct may be viewed by the faculty member as being efficient. You may find that talking with faculty outside of the classroom, where they have more time to devote specifically to you, may result in a much more engaging conversation. **For some faculty, this type of interaction may lead to opportunities that go beyond answering questions about the course, and ultimately give you the opportunity to have more meaningful discussions about areas of interest outside of the material directly being covered in the course.** Admittedly, figuring out which faculty members are open to such conversations may be a bit tricky, but taking the time to figure out the personalities of your instructors and the faculty in your department will likely help you inside the classroom as well.

In addition to the faculty with whom you are taking classes, hopefully within your program you have the opportunity to have regular advising meetings either with a faculty member or an academic advisor in your program, department, or college. These regular meetings are an excellent opportunity to gain additional knowledge about the courses you will need to take to finish your degree, as well as activities you can or should be participating in to help your professional development and career aspirations. Once your advisor understands your career goals and aspirations, she/he/they should be able to provide you with additional resources and knowledge that can help you develop an appropriate plan of action based on those goals and aspirations (see our Plan for Success section earlier in this chapter regarding the development of SEPs). This information might include specific courses to take, activities to get involved in, internships and research opportunities, and graduate school considerations, just to name a few. In addition to this knowledge, getting to know your advisors and instructors and allowing them to get to know you will be a tremendous advantage if you need a faculty or staff member to provide you a letter of recommendation for graduate school, a job referral, etc. From a personal perspective (as faculty members), we both feel it is much easier to write a letter of recommendation for someone we know well versus someone who has done well in our class, but we have little additional information about them beyond that performance. As you will see in Chapter 3 when we talk about résumés, in most circumstances when you apply for a position, there will be multiple applicants who have a similar educational background with similar success in their studies. It is the additional information about a candidate, such as their character, personality, leadership skills, and experiences outside of the classroom, that many employers will therefore focus on as a means of separating potential candidates., **Speaking toward these additional attributes will likely be much easier and more authentic for a faculty member who has worked with you as a mentor or advisor than for a faculty member who has never really interacted with you other than in the classroom.**

NETWORKING

One thing you will likely be told again and again during your academic career is how important it is to network. Many of the activities listed in this chapter are excellent starts to effective networking within your discipline. During your academic career, you will have many opportunities to meet people who are already in, have been in, or are aspiring to be in your field of interest and have discussions with them regarding career expectations and aspirations. These opportunities include activities that occur in the classroom, participating in activities sponsored by or being a part of major-related organizations, attending seminars on campus, participating in career and internship fairs, and talking with faculty and advisors, just to name a few. Many of you will find that developing your network can be as important as the knowledge you gain in your academic classes; it's not always what you know, but who(m) you know. In many ways, that saying can have terrible connotations, suggesting that incompetent individuals can be advanced to positions of authority or power simply because they are friends or family of important people within a company. That is not what we are suggesting here in this section. Instead, we are using that saying to suggest that developing your network can open doors or make you aware of opportunities that you might not otherwise have found. Your skills and abilities are what will ultimately win you the position, but you might find that your network provides you the ability to get your foot in the door.

Developing your network can be as important as the knowledge you gain in your classes, opening doors or making you aware of opportunities that you might not otherwise have found.

So, what does it mean "to network"? **Networking by definition is the exchange of information or services among different entities, be it individuals, groups, or organizations.** That definition is fairly broad, leaving the context of the information or services to be determined by not only your purpose for attending a networking event but also by what the event actually is. For example, your purposes for attending a job fair and an alumni event are likely to be quite different. Additionally, while both afford networking opportunities, the information exchanged within those events is likely to be quite different. For the job fair, you are likely attending because, as the name implies, you are interested in finding a job. As a result, the event likely will be positioned to provide opportunities for job seekers to meet with and talk to potential employers. On the other hand, you might attend an alumni event primarily to support your alma mater or perhaps to meet up with friends and colleagues. As such, the event might provide participants the opportunity to reconnect with old acquaintances and make new contacts that share a common experience (i.e., graduating from the same university). In both cases, there is an exchange of information, but the type of information and the purpose of the event are very different.

So now that we have described networking and alluded to its importance, we want to touch upon what you can do right now as an engineering student early in your career to begin to build your network. It is likely that you will have multiple opportunities to expand your network simply by going to class and engaging with your peers and faculty. These individuals will be the earliest members of your network. While many in your network (peers in particular) will be in a similar situation as you and therefore be limited in their ability to provide mentorship or advice, these individuals will gain knowledge and experience that will be different from your own knowledge and experience, and so with time, you will collectively add value to each other's networks. Beyond these initial members of your network, there are additional in-person opportunities for networking:

- outside speakers for a class or college seminar series
- student clubs and engineering societies
- volunteering at activities in your local community that have a scientific or engineering interest
- attending local, regional, or national engineering symposia or meetings

For each of these opportunities, recognize that the key word in our working definition of networking is *exchange*. This suggests that the benefit must be somewhat mutual to ensure that your networking efforts are beneficial. This will require you to understand not only how someone might be able to help you, but also how you might be able to help them as well. As a first- or second-year student, it may be difficult for you to understand how you might be able to help an engineering manager at a large firm, but these situations will likely be the exception rather than the rule. Recognize that many people in the profession will continue to "pay it forward," that is provide mentorship and advice to an engineering student much in the same way someone in their network did for them when they were a young student. Their hope would be that one day, you will do the same for the next generation of engineers.

Networking today goes well beyond the physical realm and may ultimately involve people you might never meet in person. Much like social media platforms such as Twitter, Instagram, Snapchat, Facebook, and Reddit that bring people together and develop (sometimes) virtual relationships, so do professional social networking tools such as LinkedIn, Meetup, and XING, just to name a few. Currently, LinkedIn is the de facto leader in this segment, and as such, we will refer to it when discussing professional networking tools (see Chapter 3 in regards to digital resumes where we once again discuss LinkedIn). For anyone who has used social media, LinkedIn will immediately feel somewhat familiar. Like Facebook, you can connect with people and share information via posts, although on LinkedIn these posts tend to revolve around professional activities such as upcoming events, job openings, news related to your profession, and details about personal professional accomplishments such as a new job or internship, a new certification you earned, and the like. Additionally, you can follow companies and join networks relevant to your field of study, all of which may lead to additional opportunities to "connect" with like-minded individuals. In some instances, connections of your connections (a second-degree connection) might request to connect, leading to virtual networks that extend well beyond your physically curated network. These online connections might help you become aware of opportunities outside of your hometown or college region; similarly, you might be able to connect them to opportunities you know to be local to you. In this way, your digital network can work much in the same way as your physical network does (and for many of us, there is significant overlap between the two groups); however, this digital network has the potential to reach even further if cultivated correctly.

CHAPTER SUMMARY

This chapter focuses on important factors that contribute to success in your college career. Within this chapter, the following topics were covered:

- The impact metacognition and mindset have on our success in various activities
- How to create a Student Enhancement Plan (SEP) and apply it to achieving short-, mid- and long-term goals
- Identifying professional skills including organization, time management, leadership, and teamwork and the importance of recognizing and nurturing these skills
- The value of developing a complete schedule of classes early on and the ethical considerations associated with academic integrity
- Integrating campus engagement to augment your academic career
- The importance of networking and how to begin cultivating your own network

END OF CHAPTER ACTIVITIES

Individual Activities

1. Think back on a situation in which you exhibited a growth mindset. What was the outcome of that experience? Now consider a time when you exhibited a fixed mindset. How were the results different? What made you lean more towards the growth or fixed mindset?

2. Using the process described in this chapter and the layout presented in Figure 2.1, develop your SEP for the remainder of this academic year.
3. Consider a set time period (e.g., one day, M–F, or a full seven-day week) and record how you spend your time. Once you have recorded the entire time period, look back over the summary and consider the following:
    a. What activities take up the majority of your time?
    b. Are there activities that should take up more time? Less time?
    c. What surprises you most about the time breakdown?
    d. What two or three changes (if any) do you plan to make to improve your time management, or what two or three actions do you plan to keep to maintain your time management?
4. Create a template for courses to take for the rest of your undergraduate career. Make sure to include all courses needed by your program to graduate. If you are considering more than one program, or potentially dual-majoring, create a template for each of these scenarios. Consider when select courses are offered, what prerequisites may exist, and potentially, which electives you may want to take. Does this plan allow you to graduate when you would like? How do your different scenarios compare in terms of course loads and graduation dates? What alternatives may exist?
5. Get involved with one of the engineering student organizations on campus. This could mean attending a meeting, participating in a field trip, going to a sponsored event, and so on. Submit a summary of who, when, where, and what as well as a paragraph on your experience.
6. Research and identify three to five non-engineering student organizations with which you would like to get involved. Submit a summary list of these, including contact information, their next meetings, and why they are of interest to you.
7. Conduct an in-person interview with a faculty member whom you would like to get to know better. This can be one of your current course instructors or someone with whom you may want to conduct research in the future.
8. Toward the end of the semester or quarter, look back on the SEP you developed. How are you doing in meeting goals? Are these goals still valid, and what changes might you make for coming semesters/quarters?

Team Activities

1. As individuals, review the academic conduct policy. As a group, discuss the following:
    a. Which points or expectations are the most surprising?
    b. What are the typical consequences of violating this policy at your college or university?
    c. Why is it important to adhere to this policy beyond those consequences presented?
2. Setting team expectations at the beginning of projects can ensure everyone involved is aware of and agrees to expectations. One approach sometimes taken is establishing a team contract. As a group, create a potential team contract for a semester-long project. Include items that address the following points:

    a. Contact information—who, what, how
    b. Team structure—leadership structure, how decisions are made, timelines, deadlines, etc.
    c. Meetings—when, where, agendas, minutes, attendance, etc.
    d. Participation—distribution of work, inclusivity of all members, keeping on task
    e. Accountability—expectations of individuals regarding attendance, participation, contributions to assignments, communication, etc.
    f. Consequences for breach of contract

3. As a group, identify three to five specific opportunities for you to engage in networking during this semester. For each opportunity, identify the individual with whom you would like to connect, what you would like to discuss, and the benefits that could come from the meeting for both you AND them.

CHAPTER REFERENCES

Fishman, T. 2014. "The Fundamental Values of Academic Integrity." International Center for Academic Integrity, Clemson University.

McCabe, Donald L. 2005. "Cheating among College and University Students: A North American Perspective." *International Journal for Educational Integrity* 1, no. 1. http://dx.doi.org/10.21913/ijei.v1i1.14.

Smith, Margaret, Yujie Chen, Rachel Berndtson, Kristen M Burson, and Whitney Griffin. 2017. "'Office Hours are Kind of Weird': Reclaiming a Resource to Foster Student-Faculty Interaction." Park University: *InSight: A Journal of Scholarly Teaching 12*: 14–29. http://dx.doi.org/10.46504/12201701sm.

CHAPTER 3
ENGINEERING CAREERS

BEFORE YOU BEGIN

Take a minute to consider the following questions:

- Why are you considering/did you select your current major?

- How has your past influenced the choice of your major?

- Do you have a current résumé, or have you thought about what you would include on a résumé?

- What learning experiences outside of the classroom do you hope to have before graduation?

LEARNING OBJECTIVES

By the end of this chapter, you should be able to:

- Develop a concise summary of why you are pursuing a particular degree

- Create a résumé that meets recommendations for both look and content

- Summarize critical considerations for writing a cover letter, developing a portfolio, and engaging in an interview

- Evaluate various post-graduation options and begin to formulate personal preferences for post-graduation activities

- Identify campus resources that can assist with career development and experiential learning activities

INTRODUCTION

If you are reading this chapter then presumably you are in an introductory engineering class and are thinking about, or have already chosen, an engineering discipline for your college major. Our second chapter focused on skills to develop during your academic career that will serve you well both during school and after graduation and provided suggestions for how to approach the next four years (or a little over five, if we take the national average[1]) of your undergraduate career. This chapter is going to look further into the future—with a focus on obtaining that first job or being accepted to the graduate program of your choice. It may seem a little soon to be considering your first engineering position. After all, you likely have

---

1 https://nscresearchcenter.org/signaturereport11/.

a lab due in physics or chemistry and a quiz in calculus this week. However, taking some time early on to consider where you want to be after graduation can ensure you are taking steps now on the path that will get you there with the least amount of backtracking or side trips.

Before we go too far in the future, though, let us take a minute to look briefly at your past. In the "Before You Begin" section of this chapter, we asked you to think about the path you took that got you here today and why you chose this major. While some of those reasons may be very personal (honestly, we hope, to some extent, that they are, as this enhances metacognition), we would like you to take a minute to share your experience to the extent you are comfortable.

> **Engineering in Action**
>
> Discuss with others the hows and whys that brought each of you to engineering.

Listening to your classmates should have given you a sense of the vast array of reasons people choose their major. You may have even found someone who had reasons similar to your own. Here are some of the explanations we have heard from students over the years:

- a family member is an engineer
- working or worked in a non-engineering position at an engineering firm
- always enjoyed math/science classes
- teacher/mentor suggested engineering as a career

- participated in and enjoyed field trip/camp with an engineering focus
- have military experience with engineering activities and coming back for a degree
- aptitude test suggested engineering as a career
- desire for what you can do/have with the degree

It would be impossible to summarize the numerous other reasons here. Whether you knew from an early age or only recently decided that engineering is the career for you, taking the time to formulate a clear story around these questions will come in helpful for applications (scholarship, research, or other opportunities) and possibly even internship and job interviews.

Just as each person has a unique reason for choosing engineering, we all have different ideas about what we would like to accomplish after earning an undergraduate degree. Some of you may be considering your degree as a starting point for an advanced degree in engineering, law, medicine, or the like. Others may have plans of going into business or taking a more entrepreneurial approach and starting their own. Some may be considering joining a consulting firm: either a smaller organization with a predominantly local/regional focus or a mid- or large-size firm with offices across the United States and possibly the world. Others might be looking at positions in government at a state or national level, military posts, working with a national lab, or taking a position overseas. Some may have plans to join the Peace Corps, Doctors without Borders, or other voluntary or charitable organizations. One of the greatest aspects of an engineering degree is that it keeps all of these options (plus ones we didn't mention) available to you.

One of the greatest aspects of an engineering degree is that it allows you to transition into a variety of job opportunities upon graduation.

The prospects for engineering graduates are quite promising. The Bureau of Labor Statistics projects job growth in this area of 6% between 2020 and 2030, with the majority of this job growth in the engineering occupations.[2] Additionally, engineering ranked second in the National Association of Colleges and Employers 2021 projected starting salary survey list (just behind computer sciences)[3] with a projected average starting salary for those with bachelor's degrees of approximately $71,000, a 1.6% increase from 2020 projections. (As a note, engineering is second on the list (again behind computer sciences) for master's and doctorate degrees as well, with even higher salaries: ~$82,000 and $97,000, respectively). While salaries vary across engineering disciplines, the median annual wage for the architecture and engineering occupations collectively

---

2 https://www.bls.gov/ooh/architecture-and-engineering/home.htm.

3 National Association of Colleges and Employers, *NACE Winter 2021 Salary Survey Executive Summary*, 2021.

in 2020 was significantly higher ($83,160) than the median wage for all occupations ($41,950).[4] The Bureau of Labor Statistics provides extensive information on job outlook by specific degrees, including job outlooks, salaries, opportunities, work environment, work expectations, and state-specific information.

> **Engineering in Action**
>
> Go onto the Bureau of Labor Statistics website and investigate your specific engineering major (or the majors that you are considering). Create a "top ten" list of things you learned about your job prospects that you didn't know previously.

While a promising job outlook can make finding a position after graduation easier, there are also a number of factors within your control that can ease your transition from obtaining your degree to putting it to good use. The rest of this chapter is focused on the professional documents and activities beyond earning your diploma associated with the transition to the next stage of your career. The first of these is your résumé.

RÉSUMÉS

A résumé, often called a curriculum vitae or CV in academic settings, is in many cases the first impression you provide to a prospective employer. Even in networking situations where you have had the opportunity to meet face-to-face and speak briefly with a representative of the company (we will come back to these conversations later in the chapter), your résumé is many times what is physically taken away from the encounter. Thus, it becomes the introduction to or reminder of what and who you are and is not something to be thrown together with limited thought while binge-watching your favorite show. Colleges and universities recognize this and provide resources to assist with crafting résumés and many of the other job preparation items we will discuss in this chapter. We have even included a specific section at the end highlighting what some of these resources are and where you might start to look to find them. While we encourage you to take advantage of all your campus resources (you'll see these suggestions throughout this book), we also want to provide more information here about crafting a résumé. As a starting point, let's discuss some of the broad generalities on creating a résumé.

---

[4] https://www.bls.gov/oes/current/oes_nat.htm.

> **Engineering in Action**
>
> As a group, discuss five things you feel must be on a résumé, three things you believe should not be on a résumé, and three "standard practices" for creating a résumé.

Two main aspects of a résumé are look and content. Everything else being equal, content will dominate; that is, better content will be more competitive. However, this assumes the look meets minimum criteria because there are certain things that, depending on the company, can guarantee your résumé gets placed on the "do not consider" pile regardless of content. Let us focus first on the overall look of the résumé and consider the following:

- length
- scope
- organization
- readability

Length

While the acceptable length of a résumé can vary, for the vast majority of the positions to which you are applying, a single page is the target, with a second page potentially being justifiable. Many people mistakenly believe that a longer résumé means more experience. **It is possible to craft a robust résumé, however, that conforms to page limitations if you are intentional about what is included.** This leads to the topic of scope.

Scope

Your résumé should be designed to highlight your abilities and achievements related to the description of the job for which you are applying. Résumés are never designed to tell the whole story; rather they are developed to serve as a primer for questions and conversations at the interview. The great résumés act like a movie trailer, leaving the employer with the impression you could be a good fit but wanting to find out more. The truly outstanding résumés read as if the person was tailor-made for the position. In order for this to occur, the résumé needs to be modified specifically for each job application (or job category). This is not to say you make up items to add to a résumé so you match the job description. We must always act ethically. It is NEVER acceptable to include something on a résumé that is not true, but targeting what is included is perfectly reasonable. Consider a scenario where you are applying for a position at a summer camp for elementary-aged children. Previous experience volunteering at your local boys and girls club or helping out with a YMCA or church camp would be better to include than your job as a cashier or server. You could choose to highlight skills such as CPR and first aid over your knowledge of multiple computer programs (unless, of course, this was a coding camp).

Contrast this with a résumé for a receptionist position at a local company. In this case, your computer experience, cashier or server experience, and ability to speak Spanish would be some of the key points to touch upon. It may take you a few years to gain enough experience before you need a *master résumé* which is a complete summary of all of your experience that you pull from to create a targeted résumé for a specific position. This master résumé (which is for you, not for distribution) is helpful for keeping track of your accomplishments while allowing the résumé you submit to maintain a more targeted focus.

A master résumé keeps track of all of your accomplishments and is the private document you pull from to create a public résumé targeted to the position for which you are applying.

Organization

A 2018 study[5] found recruiters scan a résumé on average for 7.4 seconds (most people probably cannot order coffee that fast), making clear organization crucial to getting a second look. **Intentional grouping of similar material, logical headings, and chronological order are all aspects of résumé organization.** Some of these considerations are ones that make sense to us, while others may require additional explanations. Grouping similar material is probably something many people would do on their own. The idea of creating separate sections for your education, work experience, and specific skills is often how this material comes to mind when thinking about items to add to our résumés. Placing these items under distinct headings (similar to how this and many other books are arranged) is the next step up in organization. Headings allow those who are skimming to more quickly locate the particular item for which they are searching, allocating more of those brief seconds of the scan to your specific talents and achievements. The final level in organization is, to some, the least logical: the use of reverse chronological order. Put your most recent achievements at the top of each section, with older material below. This is not how we think. If asked to list something, such as houses we have lived in or classes we are taking, most of us would provide information oldest to newest, first to last. But consider what the recruiter is looking to learn about you from your résumé. Is it more important that you graduated from high school or that you are currently a junior in college? Is that summer job as a lifeguard five years ago or your current part-time internship as an AutoCAD technician the work experience you want to highlight? Would you rather they know you were president of your sophomore class in high school or a current member of an engineering outreach organization on campus? In the United States, individuals tend to read from top to bottom, left to right, so organizing from newest to oldest serves as an additional time-saver; the résumé reader need only find the heading they want and look immediately below this heading to see the most current achievements.

---

5   Ladders, "Eye-tracking Study," 2018.

Readability

Linked with organization is the concept of readability. Spelling, grammar, font, white space, and more fall into this category. The spelling and grammar pieces should be obvious. No spelling or grammar errors should exist on your résumé. In addition to running the spelling and grammar check in Word or Google Docs, you should always do a second (third, fourth, …) check and have a friend read through the document as well. Computer programs are great at catching many errors but not so great for checking proper nouns (i.e., names of companies, your address) or recognizing when typos that are actual words exist in the file. The latter is analogous to texting with autocorrect, when you type "definitely" and end up with "defiantly" or try to say "life" and end up with "love." **Taking the time to have multiple real people review your résumé maximizes the chance of catching these bloopers before it is sent out.**

Font is the second aspect of readability. While we are not going to go as far as saying there is a "best" font for a résumé, we will say that certain fonts qualify as "good" ones, while others are bad, at least for this particular project. Consider Figure 3.1. The fonts on the left are the ones we consider appropriate for résumé use, while the ones on the right are much less well suited.

Times New Roman	Comic Sans MS
Arial	Corsiva
Calibri	Pacifico
Cambria	syncopate

FIGURE 3.1 Font types that are well suited (left) or poorly suited (right) for use on a résumé

Along with font type, font size should be considered as well. Many times, we are tempted to reduce the size of the font in order to fit more information on the page. While this can be reasonable to an extent, the concept reaches a point of diminishing returns (literally) in making it too difficult for a person to read. To illustrate this point, hold your book at a distance you believe most people would hold your résumé when viewing it, and look at Figure 3.2. At what point do you find yourself moving the text closer (or squinting your eyes) to read the page?

Assuming the person reading has good vision, a typically accepted minimum font size is the third line of Figure 3.2. For most fonts, this correlates to a 12-point font size, but be aware that different fonts of the same size can have different levels of clarity. While this is the minimum font size that can easily be read, we encourage you to use larger sizes for headings and subheadings in a consistent manner. In conjunction with meaningful font size variations, the use of boldface, underlining, and italicizing can also provide visual variations that complement rather than distract from your résumé.

Font size matters on a resume.

Font size matters on a resume.

Font size matters on a resume.

Font size matters on a resume.

Font size matters on a resume.

Font size matters on a resume.

FIGURE 3.2 Demonstration of various font sizes

Boldface, underlining, and italicizing can provide complementary visual variations in a résumé.

Closely linked to these font choices is white space. Aside from adjusting font, eliminating white space is what many consider an easy sacrifice. As we mentioned previously, it is tempting to fill the page by squeezing in "just one more thing." (Wow! Look how much they have done!) Rather, instead of making your résumé more impressive, it looks overly crowded and disorganized. (Wow. This hurts my eyes; look away.) As counterintuitive as it might seem, sometimes nothing (i.e., intentional white space) really is something. Just like our recommendation for a grammar or spelling check by a friend or mentor (read: human), having someone (not the computer) look over your résumé for other readability aspects like font and white space is a good investment before sending it out.

One aspect of readability that we have not mentioned is that of color. If you read through various websites that provide advice regarding crafting a "winning" résumé, you will likely come across a variety of opinions when it comes to the use of color. Some will say that color might be appropriate for certain creative positions such as graphic designers or advertising but that it should be used sparingly and in a way that the overall layout remains easy to read. Alternatively, others suggest you should shy away from the use of color, as it suggests the applicant is overcompensating or attempting to distract from a lackluster résumé. **Both camps seem to agree that a black and white résumé, perhaps with the use of grayscale in appropriate quantities, is still the most tried and true format and that applicants are best served by focusing on substance over style.**

That being said, should you choose to use color in your résumé, there are obviously some color choices that are better than others. Here again, readability is the most important factor when considering color combinations, recognizing that you might choose to change either the font color or the background color. This information will also prove useful when we examine graphical communication (part of Chapters 5 and 6), where data representation in graphs and figures and poster and PowerPoint presentations will be discussed.

When researchers looked at various color combinations and their effect on legibility, several of the tests that were conducted were done so on computer screens, which might not translate directly to printed materials. Researchers looked at legibility on LCD screens,[6] as well as combinations when projected to larger surfaces[7] and found that, in general, dark texts (black, blue, red, brown, and violet) were much more legible and more pleasing to the audience than the inverse (light text on a dark background). White backgrounds performed the best, but some combinations with yellow backgrounds (black and red) also performed well. For studies that examined printed materials,[8] similar results were found, but additional color combinations (green on white, blue on yellow) performed better than those same color combinations on computer screens. The take-home message from these studies is that you are likely better off using dark type with a white background as much as possible in your visual communications. For résumés in particular, black on white likely remains your best choice.

Content

Now that we have the look of our résumé under control, the second aspect is the content. While we alluded to some content items when we were discussing scope and organization, it bears expansion here. Even if we are crafting a targeted résumé from our master résumé, several sections or items would be expected to appear on all résumés:

- Name and contact information
- Education
- Work experience
- Skills
- Other

Name and Contact Information

The top of the page always includes your name (first and last) as well as any professional designations (e.g., PhD, MD, EIT). Below this should be your contact information, such as address, email, and phone number. If you include your address, make sure it is up to date. (We each personally had half a dozen different addresses during our undergraduate career; it is easy to forget to check this.) Your email address should be the one you check on a regular basis but should also be professional. While partyguy@... and 420lover@... are probably obvious nonprofessional choices, things like avidhunter@... or even bookworm37@... can be viewed as nonideal by some recruiters. Stick with either your school email (a .edu address is a subtle

---

[6] Iztok Humar, Mirko Gradisar, Tomaž Turk, Jure Erjavec, "The Impact of Color Combinations on the Legibility of Text Presented on LCDs," *Applied Ergonomics*.

[7] Massimo Greco et al., "On the Portability of Computer-Generated Presentations: The Effect of Text-background Color Combinations on Text Legibility," *Human Factors*.

[8] T. Nilsson, "Legibility of Colored Print," *International Encyclopedia of Ergonomics and Human Factors*.

reminder that you are a student) or a name-based address (joesmith519@...), and remember you can always set these accounts to forward to one you check more regularly.

The same suggestion of professionalism exists for any phone number you provide. We are not suggesting a completely different line or number, but answering calls (of unknown origin) with "Yo!" or a quirky but not entirely appropriate voicemail greeting while you are searching for a job or internship should be avoided. Along these lines, do not provide personal Facebook pages, Twitter handles, or the like on your professional résumé. We will talk about your professional online presence later in this chapter, but for now, think of separating personal and professional.

Education

Education is another section that is included in a résumé. If you are currently enrolled in a college or university, the name of the school, the degree(s) you are pursuing, and your anticipated graduation date are the minimum details that are required. If you are pursuing a minor, this can also be included. Your GPA can be included but is not required; you will hear differing opinions as to whether or not you should include your GPA. Consider whether your GPA is an accurate representation of your knowledge. Can you explain the value to an employer? If the explanation is more of a series of excuses, it may be best not to include it. There are other ways to highlight academic achievements, such as the dean's list or an equivalent that can be included. If you are in your first year of college, it is reasonable to include your high school information. After your freshman year, however, high school information no longer needs to be included. If you were awarded an AA (or any other degree beyond a high school diploma), this can and should be included on a permanent basis. As we discussed in the organization section, degrees are presented in reverse chronological order, meaning the oldest is on the bottom and the most recent (likely the one you are pursuing) is on the top. Another option to include in this section is relevant coursework, but it is also likely only applicable to the first year or so.

Work Experience

Like education, the general work experience category is always included, with information again being presented in reverse chronological order. This is one section where how you present the information can vary greatly and where the "look" aspects discussed previously come into play. As a starting point, you need to include your job title, the name of the company, and the dates you worked there. If you are still working, you would list the date you started and then include "– present." Below this, people are tempted to add details about their job responsibilities. While additional information is beneficial, we contend that instead of simply listing what it is you are required to do, **provide outcome-oriented statements that have broad applicability**. Table 3.1 provides a few comparisons of job responsibilities versus outcome statements for select positions.

TABLE 3.1 Comparison of Job Responsibilities versus Outcome Statements

Job Responsibility	Outcome Statement
Bartender, server	Member of two-person team providing client-based service for ten-person bar as well as twenty-five tables in a fast-paced environment serving between 450 and 500 people an evening
Mowing lawns, cleaning pools	Coordinated exterior landscaping for eight clients on a weekly basis, including scheduling, maintenance, invoicing, and payment confirmation
Cashier	Responsible for processing both cash and credit payments, ensuring customer satisfaction, balancing transactions at end of shift, and depositing cash to safe when closing

The actual outcome statements can be quite different for the same job responsibilities, which is one of the main reasons they are more valuable on a résumé. One of the other reasons they are valuable is that they provide the ability to demonstrate skills and accomplishments that are transferable to other positions. The fact that you tended bar might not seem important for that engineering internship, but client-based service experience in a fast-paced environment aligns with multiple job opportunities beyond that of serving drinks.

Engineering in Action

Think about a job you have held (or a volunteer position). Share the job title and responsibilities with the group. As a team, rephrase this to possible outcome statements. Complete this activity for a job from each member of the group.

Skills

Most résumés have a skills section, although it is not always defined as such. The heading on this one may vary, depending on the job to which you are submitting your résumé. Along with that, **this is one of the sections that is likely to change the most on a targeted résumé compared to your master résumé.** Along with a list of skills, it is often common to indicate your overall competency level for each item. For example, do you speak basic conversational Spanish, or would you consider yourself bilingual? Is your knowledge of SolidWorks a familiarity, or would you be considered proficient? We have seen skill levels indicated by words, ratings (i.e., 3/5 or even filled stars or dots), or certifications (if appropriate or used for that particular skill). It may be tempting to list every computer program you have ever opened, but both you and your potential employers will benefit from an honest and thoughtful presentation of your comfort level with the skills you list. Not having the skill level may not exclude you from consideration; fabricating a higher skill level than you actually possess is much more likely to result in a negative experience.

Other

Please, do not think that there is an actual "other" category on your résumé. This is simply for discussion purposes in recognizing that additional information may be included on a résumé beyond those categories we have already discussed. One of the first "others" that comes to mind is an objective statement. Some people will say having a clear statement of your goal at the top of a résumé is important, while others will say the objective of a résumé is pretty obvious—to get the job—and thus, the objective statement is just a waste of space. If you choose to include the statement, make sure it aligns with the goals of the position and demonstrates how your objectives fit with those of the company. Including an objective statement that clearly does not match with the position or the company is a quick way to shift your résumé to the do-not-consider pile.

Additional "other" categories can be honors and awards, student organizations, leadership, and certifications, for example. The challenge with including multiple other categories is that they begin to take up precious space on the résumé, particularly as you add more and more sections. This is not just because it is more information, but each section heading and related spacing takes up lines. Some information that you want to present can potentially be combined into fewer sections in a meaningful way. For example, you may have limited work experience but are involved in multiple student organizations where you hold various leadership positions and have participated in local, regional, or national conferences or competitions. These can be grouped under organizations and involvement, with the names of the organizations and dates as secondary headings and your positions, key events, and so on as points below each.

COVER LETTERS

With technology continuously evolving, the concept of a cover letter looks, in some ways, very different now from what it did fifty years ago. When résumés were on physical paper sent through the mail to a company address, a cover letter was a separate sheet of paper that allowed individuals to introduce themselves and highlight key points of their résumé that they felt were most applicable to the advertised job. Back then, résumés tended to be more static: the concept of a targeted résumé had not yet gained traction, so a cover letter was one of the best ways to demonstrate how past experiences related to a current job opportunity. Aspects of this are still true today. Cover letters provide you with the opportunity to speak to company or job-specific items about which you are particularly excited or the degree of alignment between your past experiences and what is expected of the position. In the same way that a résumé is targeted to the specific position, a cover letter should be written to the specific company and about the specific job.

In the same way that a résumé is targeted to a specific position, a cover letter should be written to the specific company and about the specific job.

Some of the mismatches that can occur in a cover letter are obvious: "I'm very interested in the Field Engineering I position with Company X," sent to the human resources (HR) department in Company Y. Others are subtler: "My past field experience aligns well with your advertised job expectations," when in actuality the job is an office position. Both can be addressed by paying attention to detail and being willing to set aside both time and effort in researching the company prior to applying for a position and writing the cover letter. Consider the subtler error above, where the person mentions their past field experience for the office position. Highlighting how past field experience can benefit a person in an office position or mentioning how you are interested in augmenting your existing field experience with the proposed office experience demonstrates what you can bring to the company as well as what you hope to gain from the experience.

While **the primary goal of a cover letter is to highlight your unique qualifications for the position**, mentioning how your values align with the values of the company or providing other indicators that you have researched the company beyond simply reading the job posting will often result in a cover letter that reads much more distinctly than those that are completely generic, or even generic to a particular type of position. This does not mean that a completely new cover letter needs to be developed for each position to which you apply, but it does suggest that a little extra time editing and polishing that letter for a specific company and position is worth the effort. Even if the cover letters of today are often emails sent with an attached résumé, the same care holds true for electronic communications. In addition to emails and résumés to specific companies, technology is increasingly allowing job seekers to post information about themselves in an electronic format.

ELECTRONIC RÉSUMÉS

The term electronic résumé can refer to several different things, from the computer file you developed that is then printed for a physical résumé, to any of multiple online sites that allow you to summarize and present the equivalent of your résumé. We will talk about both of these broad categories in this section. We will start with the computer file. Often this is a Word document, but we have seen PowerPoint, Publisher, and even AutoCAD used as a means of developing a résumé. As long as you are able to address the look and content sections above, the computer program you use is not particularly important. That being said, if you are sending or uploading an electronic version of this file to a prospective employer, it is often best to save it as a .pdf file, as these are almost universally accessible and display in an expected format regardless of which computer opens the file. If you have ever shared a file between a Mac and PC, you have a great idea of what can happen just opening the same document on two different machines.

Additionally, online sites such as LinkedIn allow you to post information that is equivalent to your résumé. These sites often address many of the "look" aspects discussed previously, naturally providing some polish and a fairly standardized layout to your information. Recommendations for content are the same, regardless of if you are creating a paper or electronic version of your résumé, so make sure to reference the discussion above when adding information to your site. However, there are some additional considerations in these virtual platforms that should be addressed to ensure your profile leaves a positive impression. The American Society of Civil Engineers (ASCE) interviewed a LinkedIn senior account executive[9] who provided several suggestions, such as use a professional-looking profile photo, write a meaningful headline, and craft a strong summary.

> **Engineering in Action**
>
> Go on LinkedIn and create a basic profile for yourself. Establishing an initial profile allows you to begin networking with colleagues and potential employers. Take some time to review other profiles (noticing things that jump out as effective or maybe not-so-effective) as well as investigate current job postings. We encourage you to regularly update this profile throughout your academic career.

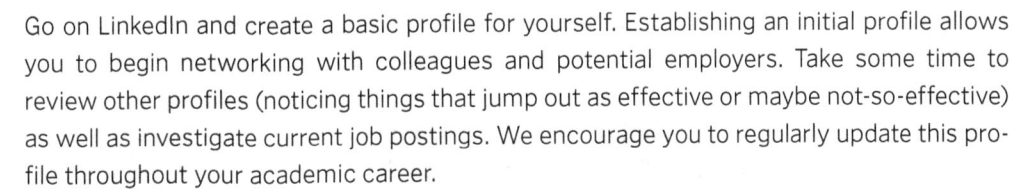

Regardless if your résumé is physical or virtual, it should contain a concise summary of your background and skills rather than extensive details. An expansion of your accomplishments can be presented in a portfolio, which some choose to reference on their résumés.

PORTFOLIOS AND ELECTRONIC PORTFOLIOS

Unlike a résumé that has a defined scope and some relatively tight space limitations, portfolios (either physical or electronic) allow for an expanded demonstration of your skills and achievements. It is easy to respond to this limitless (at least in the electronic sense) space with an approach that says everything can be included. Resist this urge. **Both content and organization are just as critical in this expanded space as they were in the development of your résumé.**

Portfolios allow for a coordinated and intentional expanded demonstration of your abilities and achievements.

---

9 https://news.asce.org/how-to-rock-your-linkedin-profile-and-land-a-job/.

The greatest value in portfolios lies in their ability to provide clear examples of skills you mention within your résumé. Your résumé states that you are bilingual; your portfolio can contain videos of you introducing yourself in French and discussing your study abroad experience in Paris last year and/or a paper you wrote for your advanced French course. You mention your proficiency in AutoCAD on your résumé; renderings of your most recent project can be included in your portfolio. Maybe you highlight your leadership skills on your résumé; your portfolio could expand on the food drive you coordinated and include the newspaper article or news clip that showcased this achievement. Remember that it is important for your portfolio to do the following:

- relate to the primary sections of your résumé (it is good to explicitly call out what you want a viewer to take away from certain files)
- contain select examples of skills and achievements (not feel like cold storage of every assignment you have ever completed in college)
- have a personal touch that highlights why you selected these particular files or documents (but maintain a professional feel—this is not your personal Facebook page)

Keep in mind that if your portfolio is in an electronic format and is public, you do not have control over who views it and when and how long they look at it. Whether you choose to make your portfolio publicly visible or provide some form of restricted access, remember a portfolio is often something that a prospective employer will look at without you being present. For this reason, providing an introduction or orientation to the portfolio ensures both that those visiting feel welcome to explore and are more likely to view or interpret specific items as you intended. Physical versions of a portfolio are more likely to be viewed for a shorter amount of time during an interview, which can allow you to add comments to the files, but you may not be comfortable carrying the equivalent of a three-inch notebook to the interview. The advantage of the physical version is that you have greater control over who views it and for how long. The benefits of an electronic version are that it can contain a wider variety of items (documents, videos, weblinks, etc.). Access to an electronic portfolio can be provided by including a link or QR code on your résumé or on a business card that you provide during an initial meeting or at the end of an interview.

INTERVIEWS

As important as it is to have a well-written, well-formatted résumé and cover letter and perhaps a robust portfolio highlighting your previous work and skills, ultimately, those elements are only designed to do one thing: get you an interview with the company for whom you want to work. When you finally get to the point of having an interview, be that in person, over the phone, or via video conference, you should recognize that your previous hard work has paid off and that the company is interested in your application. During the interview process, companies will look beyond the tangible knowledge and skills (sometimes referred to as "hard skills") you have described in your written documentation and focus as much or even more of their attention on determining your intangible qualities ("soft skills," more recently called "professional skills") to ensure that you are a good fit for the team, division, and/or company. You may find that many companies include multiple interviews during the application process, which could include a phone interview prior to an in-person interview, or multiple in-person interviews with various members of the team, management, human resources, or combinations of all three. Understanding the structure of these interviews, being prepared for what types of information the company will be looking for in the interview, and recognizing that the interview process is designed as much for you to learn more about the company as it is for the company to learn more about you are three things that will help you best represent your true self in the interview process.

Recognize that the interview process is designed as much for you to learn more about the company as it is for the company to learn more about you.

One of the first things you will need to understand about the interview process is the structure the company you are interviewing with uses for their interview process. Many companies interview in a two-step process, with the first interview occurring via phone or video conference, followed by a second in-person interview. You can think of each of these steps as a selection process, where candidates are selected on the basis of their application materials for a phone interview, and then a specific number of phone interview candidates are selected for an in-person interview. This selection process allows companies to only conduct resource-intensive (time, personnel, etc.) interviews on individuals they think are the best fit for the position, using the information gathered not only from the candidates' résumés, but also through the phone interview. The in-person interview then might consist of multiple one-on-one interviews with a variety of people from within the company or even group interviews either with multiple company representatives or even multiple job applicants. Most of this information should be conveyed to you by the prospective employer, as each company will have its own format and schedule. Our students often will ask how long it might take for a company to invite you to an in-person interview after having a phone interview, and that ultimately depends on the company, where they are in the hiring process (were you one of the first or last candidates to be phone interviewed), and how quickly they need to fill the position.

Beyond understanding the mechanics of the interview process, **it is also good to understand the company's culture so you can be as prepared as possible during the interview process.** If you are unable to determine what that culture is prior to your interview, observations such as how the employees address one another when passing through the hallways or talking to one another, how they introduce themselves to you during the interview process, and what they are wearing all can help you determine what level of formality you should express during the interview (in regard to your gestures, interactions with those interviewing you, etc.). Additionally, what types of facilities are available within the company that employees use (cafeteria, gym, child care, etc.) may help you better understand the company's expectations regarding work–life balance, team-building, and work environment. While some of these nuances may not seem important to you, to many companies, understanding how your personality fits into the company culture is a big part of ensuring that a positive work environment is maintained.

In addition to understanding the company culture, having an idea of what types of questions a company might ask during the interview is important in making a strong impression. Just like you would try to anticipate what questions a professor might ask on an exam, and just like you might prepare for an exam by working through different types of practice problems, you should follow a similar approach when it comes to interviews. Although every company will be different in its approach, in general, questions during an interview can be placed into three separate buckets: behavioral questions, situational questions, and skills- or education-based questions. We admit that this list is somewhat incomplete and that additional types of interview questions exist, but we feel these are the types of questions that most often come up during the interview process.

- *Behavioral questions* focus on things you have previously done and roles you have played in projects, jobs, and so on. Here, companies are looking to better understand your intangible qualities such as work ethic and attitude.
- *Situational questions* often revolve around common scenarios that happen in the workplace and can provide insight into your decision-making process.
- *Skills- or education-based questions* can help better define any inconsistencies a company might find on your résumé or help them better understand how well you truly understand a topic of high importance to the position. These questions might also provide insight into how well you can use the information you've gained through your education and/or previous work experience in a practical manner.

As you practice for an interview, you want to make sure that your answer is clear and concise and that it fully answers the question the interviewer is asking. In terms of how you might frame your answer, one method we recommend, particularly when it comes to behavioral questions, is the STAR response method. STAR stands for **s**ituation, **t**ask, **a**ction, and **r**esult. For behavioral questions, which typically start out with something like "Tell us about a time ...," it is important to be able to concisely describe the *situation*, explain the *task* that you were assigned, describe the *action* taken to complete the task, and conclude with the ultimate outcome or *result* of the action. In many ways, framing your answer using this method is very similar to the outcomes statements we described in the résumé section.

The last thing we encourage you to think about during the interview process is that **an interview is a two-way street: it is as much about you finding out more about the company as it is about the company finding out more about you**. A good employee/employer relationship requires acceptance from both sides, and the only way you're going to know if a company is a good fit for you is to do your research prior to the interview and ask questions during the process. Many interviews will give you an explicit opportunity to ask questions ("What questions do you have for us?" often comes up during an interview), and if this is the case, it is expected that you will take full advantage of the opportunity. That means to research the company extensively prior to your interview and come prepared with a list of questions you want to ask. Depending on what other activities are involved in the interview process (perhaps a tour of the facilities or a brief meeting with other members of the team) or what information is provided to you during the interview itself, it is likely that some of those questions will be answered, and perhaps even new ones will arise, but having a list in hand ensures that you've done your research and you will be able to impress upon the interviewer that you have given substantial thought to the position. In the unlikely scenario that all your questions have been answered, you can simply state that you had a number of questions ready coming into the interview, but they did an excellent job addressing those questions throughout the interview process.

INTERNSHIPS, CO-OPS, AND REUS

One term you're likely to encounter as you begin to think about how best to prepare yourself for your future career path is the idea of *experiential learning*. Broadly speaking, this term has come to represent any type of structured learning experience that occurs outside of the traditional classroom. All three of the experiences we have listed in this heading, internships, cooperative education programs (co-ops), and research experiences (for undergraduates) (REUs), are good examples of *experiential learning*. In addition to being a learning experience outside of the classroom, experiential learning often integrates knowledge and theory learned in the classroom with practical application, be that in a professional setting such as a business or company or in an academic or governmental research laboratory. In this section, we will talk about these experiential learning opportunities and how it is in your best interest to start considering which one(s) make the most sense for you.

Experiential learning integrates knowledge and theory learned in the classroom with practical applications.

One of the key sections of your résumé we previously discussed was the section regarding your work experience. Within this section, beyond simply keeping a record of every work position you have had, we suggested you take the opportunity to highlight specific skills and accomplishments that are either required or desired for the position for which you are applying. The more relevant that work experience is in relationship to the position, the easier it is for the potential employer to understand how those skills are transferable. **For most engineering positions, no experience is more transferable than an engineering internship or co-op.** Not only do these positions often include daily activities that might also be required of junior personnel within a company, but they are a long-term demonstration of your ability to be productive within the expectations/requirements of a company. As you get to your senior year and begin to apply for full-time positions, you will be in the same position (depending on your major) as anywhere from approximately 250 (mining) to 32,000 (mechanical) other students who will be graduating the same year as you with the same degree (in the United States).[10] Think about how many of them will be applying for the same positions as you are. They all have the same degree as you. If they are part of your graduating class, they likely will have many of the same experiences as you. How do you differentiate yourself from your peers? Engaging in activities outside of the classroom, specifically internships, co-ops, and research activities, are excellent ways of doing this.

Depending on your university, there may not be a significant difference between the term "internship" and "co-op"; some schools use these terms interchangeably, while others consider them very different. For our purposes, internships are formal fixed-term extramural experiences with a company or organization that can be either part time or full time, depending on the situation. While full-time internships typically last anywhere from eight to twelve weeks and are typically scheduled during the summer when students are often not in school, part-time internship positions might be available throughout the year and may require students to balance work and school responsibilities. Because of their timing, these types of positions are often filled by students who live in close proximity to the organization; as such, if your summer residence is somewhere other than the location of your university, you might be better off looking for a summer internship opportunity near where you will be living during that time.

Cooperative education programs share many of the same attributes as internships, and this is why at some schools, the terms are used interchangeably. Like some internships, co-op programs employ students to work within a company full time; however, unlike internship programs, these employment opportunities are for much longer periods of time, typically ranging anywhere from three to six months depending on the school (anywhere from a quarter to a semester or even a semester plus the adjoining summer term in some cases). These programs are often facilitated through the university and structured in a way such that a student is expected to complete multiple experiences prior to graduation, often with the same organization. Because these occur during the

---

10   https://www.asee.org/documents/papers-and-publications/publications/college-profiles/2018-Engineering-by-Numbers-Engineering-Statistics-UPDATED-15-July-2019.pdf

academic year, co-op students typically take an extra year to graduate but also have a year or more of work experience as a result. These programs may also focus on partnering with local industry but also might interface with companies in different parts of the country; as such, students may need to relocate for the period of time during which they are employed. Unlike some of the internship programs described above, these are almost exclusively full-time employment opportunities and most programs will require students to not take any classes during the times they are co-oping (hence why it takes an additional year or so to graduate).

> **Engineering in Action**
>
> Go to a job listing website (such as Indeed.com) or a local company's website (either local to your university or where you will be living for the summer) and look for internship opportunities that exist related to your field of study. Can you find at least three internships that sound interesting to you? If your school has a cooperative education program, what are the requirements to enter that program, and what companies participate in that program that are relevant to your field of study?

Unlike internships or co-ops, which typically are run in conjunction with industrial partners (private or public companies, government agencies, etc.), research experiences are typically run through educational entities, either your home institution or another institution of higher learning. Some research experiences also exist within government agencies as well, such as the National Institute of Standards and Technology (NIST) or the National Institutes of Health (NIH), and within some of these agencies, the line between research experience and internship might be blurred. Although these programs might have a variety of different names, including summer undergraduate research fellowships (NIST) and summer internship programs (NIH), we have lumped these research opportunities under research experiences for undergraduates (REUs). Our distinguishing factor in separating these experiences from internships and co-ops is that they are typically more focused on the scientific process associated with discovery compared to research that might be conducted in an industrial setting that is focused on product development. **For students who have an interest in a career in academia or are looking to apply for admission to a postgraduate program in engineering, having a significant research experience on your résumé is just as meaningful as an internship or co-op is to someone looking to enter industry.** Research experiences also focus on many of the same soft skills as internships and co-ops do, so many students will benefit from them even if they are not interested in academia or postgraduate education.

We use the term REU to encompass all research experiences, and for some of you, that may leave you scratching your head. For those who have never heard the term REU before, REU, as we have suggested, stands for **R**esearch **E**xperiences for **U**ndergraduates. An REU program, though, is a program run either by a governmental agency or university sponsored by the National Science Foundation (NSF) that supports active research participation by undergraduate students in a variety of research areas. These programs typically consist of approximately ten students per site and provide undergraduate students the opportunity to work closely with faculty and other researchers on projects[11], much the same way an internship or co-op would work. These paid opportunities occur across the United States and typically cover costs associated with housing and travel. Each program is unique in its area of research, as well as the type of students they are looking for. You can search for REU opportunities through NSF's website.[12]

Recognize that most research experiences that undergraduates participate in will not be REU programs sponsored by NSF. Many such opportunities might exist at your current institution, either within your program of study or in other programs across campus. Depending on your situation, you might also find opportunities available at other institutions as well. Unlike internships or co-ops, these opportunities may not be heavily advertised or advertised at all, and there may not be a central location such as a job-aggregating website or a jobs portal within a company's website for you to view available opportunities. Additionally, you may find that opportunities that are available are not paid positions, similar to some internship positions. If you are looking for opportunities within your home institution, our suggestion is to figure out what resources you have available on your campus that can help you connect with the right people and organizations. In addition to these resources, talking with your advisors, student organizations associated with your professional engineering organizations, and faculty within your home department are also good opportunities to gain a better understanding of what exists within your university.

Engineering in Action

Go to the NSF REU search page and look for research opportunities that exist related to your field of study. Can you find at least three different programs that sound interesting to you? You might need to examine their individual websites (separate from the NSF website) to better understand the projects that students have worked on in the past.

In addition to serving as relevant work experience on your résumé, experiential learning opportunities often provide insight on companies and activities of particular interest (or lack of interest) to you moving forward. **Learning both what you are interested in pursuing as well as**

---

11 https://www.nsf.gov/crssprgm/reu/.

12 https://www.nsf.gov/crssprgm/reu/reu_search.jsp.

what may be of significantly less interest to you at this time are both valuable aspects of the experiential learning activity. These activities can also assist in formulating your thoughts on postgraduate studies.

GRADUATE SCHOOL

From our experiences, there are two different types of students that are thinking about graduate school (as well as another group with no intention of going to graduate school, which is also just fine), and as such, their questions and the information they are looking for are very different. We will do our best in this section to address both groups that are interested in knowing more about graduate school while recognizing that specific programs or majors might have additional considerations when it comes to graduate school. For the purposes of this book, we will consider graduate work to be anything beyond your undergraduate engineering degree (i.e., it does not need to be a graduate engineering degree). This would include administrative and management degrees such as a master's degree in business administration (MBA) or a master's degree in engineering management, professional degrees such as a doctor of medicine or a juris doctor, or even master's or doctoral programs in disciplines other than engineering, such as chemistry or physics.

Every year we see a cohort of students whose next step in their professional path is an advanced degree either in engineering or some other advanced topic. These students typically know early on that their goal is perhaps a job in academia or in a profession such as medicine or law, where your undergraduate degree is part of the foundation you build prior to being admitted to a graduate program. For these students, graduate school questions typically revolve around the application process for these various programs, the requirements of getting accepted into a program, and gaining a better understanding of which graduate program within their chosen field of study (engineering, medicine, etc.) is the best fit for them. Obviously, this is something that will vary greatly depending on the type of program you are interested in, and as such, providing additional information would be a book in and of itself. The advice we typically give students in this category is as follows:

1. **Be aware of your timeline.** The earlier in your undergraduate career you recognize that your professional path will include graduate school, particularly if that schooling is to occur immediately after the completion of your undergraduate degree, the more time you have to prepare for the process.
    a. Do the programs you are looking to apply to require you to take a graduate admissions test?
    This will require some research on your part, as this varies both by type of program as well as the institution. The GRE is the most common exam for a technical graduate program, although some programs accept an FE instead, while the GMAT is most common for management, LSAT for law, and MCAT for medical school. More recently, programs are relying less on standardized tests, so this may not be a requirement for

some of the programs you are considering, hence the importance of looking at each program individually.

    b. If tests are required, when should you take them? Are there specific courses you should take prior to the exam to ensure you have been exposed to those materials?
Although the timing may vary depending on which exam you need to take, these are typically taken no later than the summer prior to your senior year of college if you want to be admitted to your graduate program the summer or fall after you graduate. Many students will take them earlier than this, giving them the opportunity to retake the exam should they not score as well as they had hoped. Depending on the type of test you are taking, you might consider taking courses in a specific sequence (such as Organic Chemistry or Biochemistry for students planning on taking the MCAT) so that you have taken those courses prior to taking the test, if possible.

    c. What are the additional requirements for your application? If you need a letter of recommendation, when should you ask for one?
We recommend referring back to Chapter 2 in regard to faculty and advisor relationships, which are critical for obtaining a high-quality letter of recommendation. In regard to timing, we recommend giving faculty at least six weeks if at all possible, recognizing that they may have multiple requests for letters of recommendation that have similar deadlines to yours. Make sure to look closely at application due dates. While many are accepted on a rolling basis, often there are set deadlines (often early) to be considered for scholarships or funded positions within the program.

2. **Recognize how the individual components of the application are judged by the admissions committee.**
Although each type of advanced degree program we have previously mentioned is somewhat different, and each application will have different requirements, here are some general pieces of advice: ensure your GPA is appropriate for the type of program you are interested in pursuing, verify you have taken the appropriate coursework necessary for the graduate program, (particularly for professional programs in various areas of medicine such as medical school, dental school, or pharmacy school), and confirm your extracurricular activities appropriately demonstrate your commitment to the field (be that volunteering, community involvement, leadership, research experiences, etc.).

A second cohort of students we see each year intend to find a job in industry after graduation, but they are not necessarily sure if or how graduate school might fit into those plans. Here are some questions we repeatedly see from students:

- Do I need an advanced degree?
- What type of advanced degree should I get?
- Do I need to go to graduate school immediately, or can I wait until I get a job?
- Will my employer pay for my graduate school?

Obviously, the answers to each of these questions will depend on your individual circumstances, including your field of study, the nature of the position you are applying for, and the type of company you ultimately work for, just to name a few. An exercise we often encourage our students to do is to look for positions in their field of interest that would be of interest to them at different points in their career trajectory. In other words, not only look for jobs you would be qualified for immediately after you obtain your degree, but also look for jobs you think you might be interested in five to ten years into your career. When you look at those non-entry level positions, what requirements do they have? If you see experience is the thing they are looking for, that suggests perhaps an advanced degree is not necessary or is not as important as getting your foot in the door. If an advanced degree is necessary, that suggests that you will likely need to consider continuing your education at some point in time. This may not tell you *when* you need to continue your education but will give you a sense as to *if* you need to continue. One last thing to consider if you are weighing the pros and cons of going immediately into industry versus continuing your education; there is nothing that says you can only do one or the other. We've seen several of our students apply to graduate schools as they are also applying for positions within companies. Doing this gives some additional flexibility in determining what the best option for him/her/them is in moving forward with his/her/their career.

CAMPUS RESOURCES

At this point, you may be feeling overwhelmed thinking about résumés, internships, graduate school, and more, particularly on top of classes and everything else going on in the semester. Fear not; we are here to let you know you are not alone, and you are not the first person to wish all this "adulting" could be just a little easier. Colleges and universities recognize that while some students can be very successful on their own, there are significant benefits to providing a support structure accessible to all students (and at some schools to recent alumni as well). For this reason, **most schools have an office or group of offices dedicated to career development.** These organizations are staffed by professionals, many of whom have prior experience in HR at different companies, with review boards for graduate school or nationally competitive scholarship programs, or other relevant backgrounds, and they offer resources and support for the following:

- Résumé building (workshops, drop-in sessions, one-on-one meetings, example résumés)
- Electronic postings (LinkedIn profile, electronic portfolios, etc.)
- Interviewing (lists of questions, common mistakes, questions to ask, mock interview sessions)
- Job and internship fairs
- Test prep for GRE, MCAT, LSAT, etc.
- Competitive scholarship, graduate school, medical school preparation

Some even offer etiquette dinners and "dress for success" sessions with coupons from local stores to purchase interview clothing. These offices have been expanded over the years to address not only the needs of graduating seniors but the entire student body, from individuals just starting their first year to postgraduate students.

> **Engineering in Action**
>
> Locate your college or university's career development office. Note the specific office name may vary, but they most likely have both a physical office as well as an online presence that can be found by typing "career services" or "career development" into the search engine on your school's main home page. Take a few minutes to look at what services and resources they offer and consider making an appointment to ask for additional information about what you should be doing at this stage of your academic career.

As with everything else we have mentioned in this chapter, getting familiar with the resources offered by your school's career development office early in your academic career is more beneficial than waiting until the semester prior to graduation. Interview skills that work for your first job are also applicable as you interview for an internship or research position. Crafting a résumé early lets you hone that master résumé and practice developing those targeted résumés. Developing an electronic portfolio in your first year allows you to demonstrate growth throughout your college career. Campus resources are designed both to support your development and to provide a safe space for early learning attempts in a low-stakes environment. Taking advantage of these activities and more improves your professionalism earlier in your career, which increases the chances for you to stand out from the crowd.

CHAPTER SUMMARY

This chapter focuses on preparations and experiences outside the classroom that contribute to reaching your personal goals after graduation. Within this chapter, the following topics were covered:

- Recommendations for developing a résumé, particularly the creation of a *master résumé*, and writing a persuasive cover letter
- The importance of both look and content in résumés, cover letters, and portfolios
- Tips to help your résumé (physical or electronic) and cover letter stand out from others as well as how to make a positive impression during an interview

- Approaches for the development of a physical or electronic portfolio
- Suggestions on *experiential learning* activities to pursue while in school and how these contribute to increased opportunities upon graduation
- Items to consider if an additional degree is part of your post-graduation plans
- Campus resources available to you that can assist with many of the items and activities discussed throughout the rest of the chapter

END OF CHAPTER ACTIVITIES

Individual Activities

1. Consider what brought you to engineering as well as what keeps you interested in engineering.
   a. Summarize two or three key facts that serve as a starting point for this topic.
   b. Create a short (about one minute) answer to this question that would be appropriate for an interview setting.
   c. Develop a written response (~250 words) about why you selected your major.
2. Revisit the Bureau of Labor Statistics website and search for your current major/major of interest.
   a. Under the "How to become one" tab, read through the "important qualities" and identify specific activities that you can complete during your academic career to strengthen three of these qualities.
   b. Click on the "State & Area Data" tab and follow the link to the "occupational employment statistics" for your major. Which states or regions have the greatest opportunities? What do the opportunities look like in your current state? How about in the state in which you would like to live after graduation?
   c. Click on the "State & Area Data" tab and follow the link to the "occupational employment statistics" for your major. Scroll to the annual mean wages by state or by area graphs. Where are some of the highest wages found? What do the wages look like in your current location? What else needs to be considered when looking at these variations?
3. Create your *master résumé* and determine a permanent location for this file that can be accessed throughout your college career.
4. Go onto LinkedIn and look at profiles from multiple individuals. Identify three profiles that you believe are the most professional, and list five reasons why you believe this to be the case.
5. Return to the basic LinkedIn profile you created when completing the *Engineering in Action* activity earlier in the chapter. Update or complete this profile to the point you would be comfortable having a prospective employer review it for a possible job interview.
6. Search for and locate a co-op, internship, or research experience for which you would be interested in applying. Write a cover letter for this position following the expectations summarized in this chapter.

7. Attend an event offered by your college or university's career development office. Write a short (~150 word) summary of the event and list three important takeaways from your participation.

Team Activities

1. Have team members share their résumé with another team member. Review your team member's résumé, providing strengths and areas for improvement. Make sure to connect your comments to the criteria mentioned in this chapter.
2. Use your results from question 4 above to view at least one of the favorite profiles from each team member. Compile the individual strengths each person identified into a "Top 5" list of what the team believes to be aspects of a strong professional profile.
3. Using the "Top 5" list the team created in question 2 of team activities, view each team member's LinkedIn profile and suggest three changes that would have the biggest positive impact on each individual's profile.
4. As individuals, research the most common interview questions and select three to five potential questions. As a group, take turns asking these questions to different group members (make sure to rotate so everyone has an equal opportunity to ask and answer questions). After each response, as a group, share the strengths the responder exhibited and make suggestions for what could have been done differently.

CHAPTER REFERENCES

2Ladders. 2018. "Eye-tracking Study." Accessed June 30, 2020. http://go.theladders.com/rs/539-NBG-120/images/EyeTracking-Study.pdf.

Greco, Massimo, Natale Stucchi, Daniele Zavagno, and Barbara Marino. 2008. "On the Portability of Computer-Generated Presentations: The Effect of Text-Background Color Combinations on Text Legibility." *Human Factors* 50, no. 5 (October): 821–33. https://doi.org/10.1518/001872008X354156.

Humar, Iztok, Mirko Gradisar, Tomaž Turk, and Jure Erjavec. 2014. "The impact of color combinations on the legibility of text presented on LCDs." *Applied Ergonomics* 45, no. 6 (November): 1510–17. https://www.sciencedirect.com/science/article/pii/S0003687014000696.

National Association of Colleges and Employers. 2021. *NACE Winter 2021 Salary Survey Executive Summary*. Accessed April 8, 2022. https://www.naceweb.org/uploadedfiles/files/2021/publication/executive-summary/2021-nace-salary-survey-winter-executive-summary.pdf.

Nilsson, T. 2006. "Legibility of Colored Print." *International Encyclopedia of Ergonomics and Human Factors,1,* 1440–1452. CRC Press.

CHAPTER 4
INFORMATION LITERACY

BEFORE YOU BEGIN

Take a minute to consider the following questions:

- What do you think of when you hear the term *information literacy*?
- How much experience do you currently have with searching for technical information?
- When was the last time you were in your college or university's library? On the library's website?
- How comfortable are you synthesizing data (i.e., developing new ideas or discussions based on evidence found in multiple sources)?

LEARNING OBJECTIVES

By the end of this chapter, you should be able to:

- Define information literacy and explain the importance of being informationally literate
- Discriminate between different sources of information and determine their relative value
- Develop appropriate strategies to locate relevant information
- Conduct a systematic search for information on an engineering topic
- Synthesize information from multiple sources in a logical and cohesive manner
- Identify and avoid potential plagiarism violations

INTRODUCTION

The term *information literacy* is one you have probably heard, and you may have already taken classes that had an expectation or goal of developing your information literacy skills. Before we delve into suggestions for strengthening these skills or even why they are important, we would like to present some background on information literacy as well as the definition we will use to guide the development of this chapter. Although different organizations and resources use slightly different language to define information literacy, they all focus on the skills associated with understanding the need for information to appropriately solve a problem or make a decision as well as the subsequent ability to find, evaluate, and use that needed information effectively.[1] In a

---

[1] Presidential Committee Information Literacy: Final Report, *American Library Association*.

nutshell, an informationally literate person knows when they need information, and they know how to find and appropriately use it.

An informationally literate person knows when they need information, and they know how to find and appropriately use it.

The term *information literacy* was formally integrated into the educational experience in the early 1990s.[2] Around this same time, the Association of College and Research Libraries (ACRL) developed the Information Literacy Competency Standards for Higher Education.[3] In 2009, the Association of American Colleges and Universities (AAC&U) published a series of rubrics[4] designed to provide a measure of standardization in the assessment of select competencies. This project was called VALUE (Valid Assessment of Learning in Undergraduate Education) and included sixteen topics, one of which was information literacy. Most recently, in 2015, the ACRL released the Framework for Information Literacy,[5] which modernized and refocused earlier work to reflect current technological advances. The six main components of this framework are illustrated in Figure 4.1. As we continue in this chapter, we will focus on three of these aspects: strategic searching, research as inquiry, and the value of information.

---

2 Sharon A. Weiner and Lana W. Jackman, *The National Forum on Information Literacy, Inc.*

3 Information Literacy Competency Standards for Higher Education – The Association of College and Research Libraries (ACRL), Information Literacy, 2005.

4 Terrel L. Rhodes, *Assessing Outcomes and Improving Achievement: Tips and Tools for Using Rubrics*, Association of American Colleges.

5 Association of College & Research Libraries, *Framework for Information Literacy for Higher Education* (Chicago, 2015).

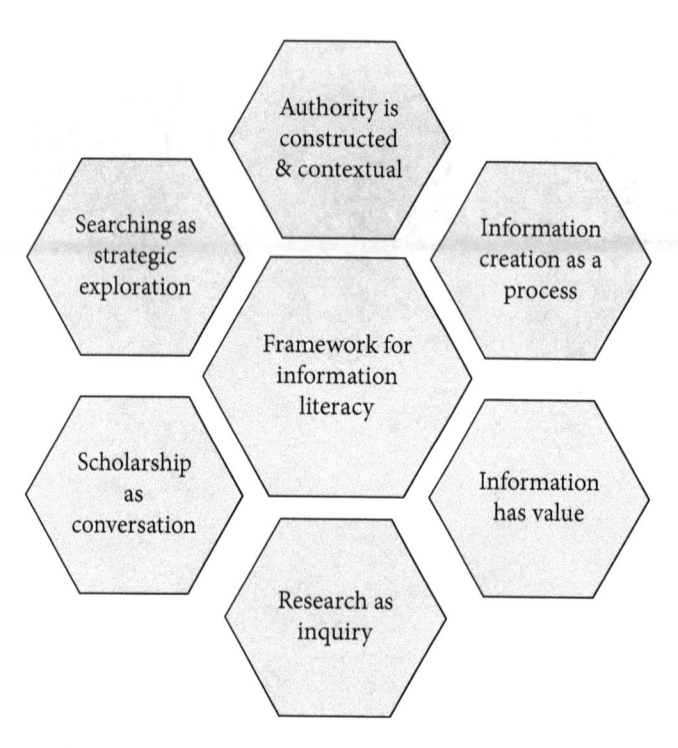

FIGURE 4.1 Key components of the ACRL's Framework for Information Literacy

In today's world, ready access to information borders on being a basic necessity of life. In the United States and other developed countries, and even in many developing countries and remote parts of the world, access to the internet is easily available. Unanswered questions may no longer require lengthy searches or reliance on the memory of others but rather can be answered with the press of a button or simply by calling Alexa. These approaches are ideal for questions such as who won Super Bowl XX (the Bears) or what happened in season eight of Game of Thrones but may be less appropriate when the questions are more complex or possess less defined responses. Part of mastering information literacy is developing an understanding of which searches are appropriate for which situations.

Strategic searching starts with the understanding of which resource(s) to use to conduct your search. Additionally, having familiarity with common methods for modifying search terms and how to expand or limit existing searches can dramatically decrease the time it takes to successfully complete a search. Treating research as inquiry can encourage persistence in your search. While it can be frustrating when initial efforts do not produce the immediate results you desire, it is critical to recognize the nonlinear process necessary to search for answers to complex questions. Many times, a comprehensive search requires initially locating potential resources, evaluating the value of these sources, refining the search based on that evaluation, and repeating this process until enough information is obtained.

> **Engineering in Action**
>
> Discuss with others the background each of you has with information literacy activities. Were these experiences in high school and/or college? Was formal training involved, or were only basic directions provided? Who has seen or utilized the ACRL Framework in the past?

As you progress through your engineering career, there will be several opportunities for you to flex your information literacy muscles. For example, you may be asked to:

- compare your results to published data on the same activity;
- investigate possible lab procedures that can test a hypothesis you have developed;
- conduct a literature review to situate your project in the larger scope of a given topic; and/or
- determine which codes or regulations apply to a proposed design.

Your searches may require historical data, current publications, or proposed future legislation, which may place restrictions on a project in the early stages of design. The need for information literacy goes beyond the idea of standing on the shoulders of giants (Sir Isaac Newton)[6] or recognizing that without the knowledge of history, we are condemned to repeat it (George Santayana)[7] by recognizing that we cannot engineer in a vacuum. Information is required as inputs for even the simplest of problems we will solve (e.g., what is the gravitational constant here on Earth), the solutions will be based upon equations and models previously defined and validated by others, and our rationale for using this information will be based on our current understanding of the problem at hand. Knowing where to find the information you need, how to use it, explain why it is useful, and describe how your solution fits into the larger context of the field you are working in is something you will do throughout your career in engineering. Developing these tools is something your engineering degree will do, even if you do not have a class specifically dedicated to information literacy.

---

[6] http://www.bbc.co.uk/worldservice/learningenglish/movingwords/shortlist/newton.shtml.

[7] https://www.iep.utm.edu/santayan/.

CURRENT RESOURCES

So now you have a working definition of information literacy, a little background as to where and how information literacy came to be within the context of education, and why it is important, both as a student and ultimately in your career as an engineer. But where do you go to find information? Today, the term "Google" (i.e., "Googling" a topic) has become synonymous with web searches, much the same way Kleenex is commonly used as a generic term for facial tissues and JumboTron is used generically to describe any gigantic TV screen in arenas and stadiums, regardless of the manufacturer. With the rise of personal assistants (Alexa, Google Assistant, Siri, Cortana, etc.), even Googling a topic has perhaps become somewhat dated. With all this technology at your fingertips, access is not the issue, and resources are plentiful. Before we consider this challenge resolved, we need to consider a few questions:

- Is Google or Alexa the best tool for the job?
- How do you know the results of your search query are appropriate?
- Are there better tools available for engineering research?

Using the Right Tool for the Right Job

General Searches—Search Engines and Personal Assistants

Using the right tool to help find appropriate information can be the difference between finding a resource that is suitable to help defend your argument or aid in solving your problem or finding a resource that is less than ideal. Personal assistants and search engines (Google, Bing, Yahoo, etc.) are great at answering well-defined questions and problems, such as, "Who was the first man on the moon?" (Commander Neil Armstrong) or "How many ounces are in a gallon?" (128 US fluid ounces). They can also easily point to appropriate resources for current events, such as today's stock market indices, who won the basketball game, or what happened on the last episode of your favorite TV show. But do their results for more sophisticated, obscure, or ambiguous queries, such as "What is the history of solar roadways," contain meaningful information that will help you appropriately analyze and respond to a problem or question? How can you ensure that the information you are provided from a search query is appropriate and meaningful (thinking back to Figure 4.1, this is what is meant by the term "authority")?

> **Engineering in Action**
>
> Enter the search term "history of solar roadways" (without the quotes) and take a look at the generated results. If you were looking for information about the development timeline for the use of solar technologies on or within roadways worldwide, was this search successful? How might you modify this search such that the outcomes correlate more directly with the intent of your initial query?

One of the potential problems with search engines is that they act much in the same way as a utility knife does; in other words, there are design compromises that must be made to make the tool as versatile as possible. When you use a search engine or personal assistant for a search, the algorithms used for the query results are attempting to interpret your intent or what you mean by your phrase. Let us go back to that example of "history of solar roadways." When we searched Google with that query, we returned 250 million results (that is a lot; more on that later), with several of the top hits referring to the home website and news articles about a company named Solar Roadways that is currently testing products associated with solar roadways here in the United States. Google did not know from our search terms that what we meant by our search was the history of research associated with using solar-powered technologies in roadways; instead, it interpreted our query to mean the history of the Solar Roadways company. So this failure was our fault; our search terms were ambiguous. We could have easily refined our search with different terms to get better results. But if this were the first time we were searching for information in this field, would we have known what additional terms to use? Furthermore, sometimes search engines (and personal assistants) just get things wrong. Because Google has knowledge of over 130 trillion unique pages (this was the number they last reported back in 2016[8]), a proprietary search algorithm is used to provide what it deems to be the most relevant results. **There is no human curation to check for accuracy or validity; you are on your own to determine if the search results are appropriate for your query.**

For most common uses, search engines are great at finding the information you need. As an initial dive into a particular problem, search engines might provide you with information that will allow you to further refine your search parameters or perhaps give a strong overview of the topic or problem you are researching. Depending on your search parameters, you might even find multiple resources that are significant in their findings and are appropriate as cited references in your work. Alternatively, you might find resources that are completely inaccurate, either regarding your search parameters or possibly even containing incorrect or non-factual content. Remember, search engines and personal assistants are not curated by a human. Furthermore, **information published on the internet does not require any type of peer or editorial review before the content is indexed by the search engine.** Because of these issues, you need to ask yourself if your search results are appropriate and reliable. In essence, you become the human curator of your own searches, filtering out the good from the bad and determining the relative level of authority of the source. This can take a lot of time and effort and ultimately may be more work than you want or are able to do (particularly if you are a novice in the area you are researching). Instead of relying on an algorithm developed by a company to determine what you are looking for, how can you gain better control of your searches?

---

[8] Barry Schwartz, "Google's Search Knows about over 130 Trillion Pages," *Search Engine Land*.

Advanced Searches and Boolean Operators

One way to take control of search results is to use Boolean operators in your search terms to focus your query. Using the three Boolean operators "AND" "OR" and "NOT" you can link specific terms and phrases in a way that provides additional specificity in your search. Although for the purpose of this section we will focus on using Google, these tips and tricks will work with any search engine (although the syntax might be different). By using "+" (equivalent to the Boolean operator AND) in front of a word or quotes around a specific phrase, Google will include that specific language within its generated results. Similarly, using "-" (equivalent to the Boolean operator NOT) in front of a specific word or phrase will exclude that information from the search result. Lastly, using "OR" between two words tells Google that pages may contain one word or the other but do not need to contain both.

> **Engineering in Action**
>
> Conduct a search using the terms "Olympics," "1896," and "2018" (all without the quotes). How do the search results change when you use the Boolean operator OR between 1896 and 2018? If you require the search to include or exclude the word "Athens," how do your search results change?

Beyond the use of Boolean operators, search engines often have additional means of filtering or specifying search results. For example, searches can be limited to specific websites or classification of websites; for Google, this is done by using the "site:" function. For our Olympic example (see **_Engineering in Action_** previously), searching for "Olympics 1986 OR 2018 site:.edu" will only provide results that are located on pages hosted by educational domains, those that end in ".edu." Often, searches using these additional functions can be accessed through a search engine's advanced search page, which will include options for Boolean operators, limit results to specific web pages or domains, and more. For Google, this can be found at https://www.google.com/advanced_search.

Topic-Specific Search Platforms

Another way Google and other platforms have tried to overcome limitations associated with the one size fits all approach to search engines is to have search engines that are more specialized, that is, they search only a subset of the internet to provide a specific genre of results. Examples of this approach that would likely be beneficial for engineering students include:

- Google Scholar (https://scholar.google.com)
- Google Patents (https://patents.google.com)
- Semantic Scholar (https://www.semanticscholar.org)

As their names suggest, these sites provide results associated with your query that are of a specific genre. For example, Google Scholar focuses primarily on scholarly materials, including those found through the websites of academic publishers, professional societies (e.g., conference proceedings), and university websites (e.g., research theses). Google Patents, as the name implies, focuses on published and pending patents from all over the world. Semantic Scholar, although it is not a direct descendent of a more mainstream search engine, uses artificial intelligence (AI) to generate appropriate search results within the genre of scientific literature (as opposed to human curation, which we will touch upon later). The benefit of such sites is that by limiting the scope of your search, you are more likely to find information that is related (and ultimately valuable) to your search parameters. Like we discussed in the previous section, these websites also have advanced search and filtering capabilities, which should be used to narrow down your results to those that are the most pertinent.

> **Engineering in Action**
>
> Based on the material presented in this section, how do you think your solar roadways search might change when conducted on a topic-specific search platform? Enter the search term "history of solar roadways" (without the quotes) and look at the generated results. How do the search results change using Google Scholar, Google Patents, or Semantic Scholar?

Going back to our history of solar roadways example from earlier in this section, running the same query in Google Scholar results in search results that look very different from the standard Google page. First and foremost, the number of results has been dramatically reduced, from 250 million to just 14,800! Not that 14,800 is a reasonable number of results to look through, but it is certainly more palatable than 250 million. This number includes results that are either citations or patents; in removing patents (if that was not the focus or interest of your search), the number of results reduces to 14,300.

In addition to reducing the number of results, you should also recognize a significant difference in the types of results Google Scholar provides. A vast majority of the top results from our original Google search were

- websites, including a Wikipedia entry, the home page of the Solar Roadways company, and one of their Indiegogo campaigns from 2014,
- multiple pages discussing the success (and failures) of the Solar Roadways company, and
- a solar roads presentation from a student group at The Ohio State University.

None of these results were **peer-reviewed material**, that is, work that has been reviewed by external content experts who can provide an unbiased opinion as to the validity and appropriateness of the material being presented. While it is likely that some websites are reviewed by different individuals (particularly large corporate websites), editorial and/or legal review of a website is very different from peer review. For personal websites, it is quite likely that even this level of scrutiny is not conducted.

Peer-reviewed material is work that has been reviewed by external content experts who can provide an unbiased opinion as to the validity and appropriateness of the material being presented.

In comparing these results to those found in Google Scholar, you should be able to quickly notice, even without clicking on any of the results, that both the focus and the content of the Google Scholar results are very different. For example, only one of the first ten results in Google Scholar is directly referencing an HTML webpage; all the others are linking either directly to a document or a record of a document (more on records below). There are also links along the right-hand side of the page that are directly linked to resources that are available to you. Within these results, you can find:

- references to peer-reviewed journal articles,
- a reference handbook on solar energy, and
- additional manuscripts from different Universities (including student research articles).

In terms of authority and focus, these results are significantly better than what we previously found using Google; however, they still suffer from some of the same issues we described for general search engines. For example, **even though these search results are more focused, they are neither curated nor authenticated**. Furthermore, just like we saw for searches conducted via Google or Bing, results are derived from the website's search algorithm; therefore, you are relying on a website developer to tell you which results are the most important. In many instances, ranking is a function of popularity. Consequently, it is still up to you to curate and evaluate your results, determining what is valid and appropriate.

Library Catalogs and Journal Databases

Beyond using search engines, your university library is an excellent source of information and has many additional options to aid in your research. In addition to the traditional library catalog, which lists all the materials the library has in its possession (books, journals, electronic journal subscriptions, videos, etc.), your library also has access to a variety of journal databases that provide significantly different results than those found through Google or Yahoo. Due to how these databases are distributed (more on that below), your university is likely to have slightly different databases available to you than we do at our home institution. As we describe the benefits and weaknesses of these platforms in the subsequent paragraphs of this section, we encourage you to talk to a librarian at your university to figure out what databases are available to you.

For both library catalogs and journal databases, you will need to take a slightly different approach to searching than you did when using a search engine. This is due to how information is collected in catalogs and databases versus search engines. For search engines, information is often collected on the entire entry, meaning that the search engine indexes everything about the web page: title, text, metadata, etc. When you search for words or a phrase, the search engine provides results for every page that contains those words or phrases anywhere within the page (although with an advanced search, you can limit where in the page those terms are found). From our earlier example for solar roadways, remember that Google produced 250 million results. Google Scholar was significantly more focused, producing 14,800 results. Unfortunately, to go through all those entries would be nearly impossible, not just because the number of results is staggering, but also because some of those sites no longer exist, or they have moved or significantly changed since they were last indexed by Google. Furthermore, some of the documents the results point to (or even the websites themselves) are behind a paywall or protected in some way that prevents you from having access to the information.

> ### Engineering in Action
>
> Enter the search term "history of solar roadways" (without the quotes) into your university's library catalog and look at the generated results. How do these search results compare to the results you generated using either a general or topic-specific search engine?

Compare the results from the Google and Google Scholar searches to those we obtained when searching our university's library catalog (Figure 4.2). We are back up to over 148,000 matching items! How can this be? First and foremost, remember that the library contains not only academic materials but also materials from mainstream media and news outlets. Therefore, these results are broader than the scope of our Google Scholar search but much more refined than that of our previous Google search. The results shown here represent materials available at our university that contain the words *history*, *solar*, and *roadways* somewhere in the document record. Of the first ten results, when sorted by relevance, we find six articles and four books. As we were able to do in previous searches, we can further refine our search to give us more precise matches to the intention of our query. One advantage of using the catalog search is that we can narrow our results to articles from scholarly publications, something we were not able to do with

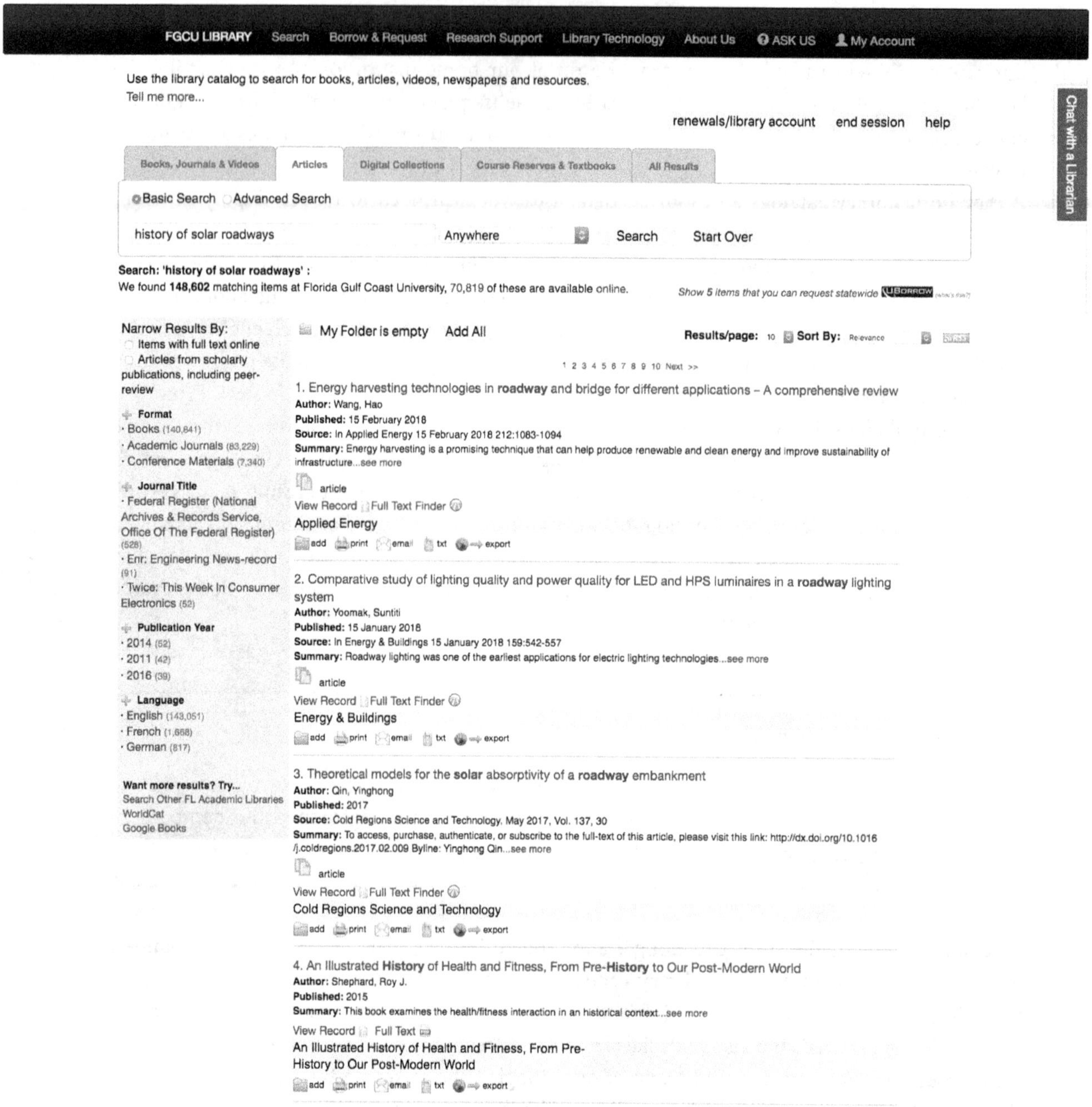

FIGURE 4.2 Results from university library catalog (Florida Gulf Coast University) search for "history of solar roadways"

90   FOUNDATIONS OF ENGINEERING

our previous searches. When we do this with our history of solar roadways search, we reduce the number of matching items to 1,980!

One of the main differences when looking at a library catalog or database search versus a search engine is understanding how the search term(s) relate to the results. As we described previously, **search engines index the entire result**, meaning that if the word(s) or phrase(s) entered in the query appear within that indexed text, the result is shown within the search engine. Compare that to a catalog or database, where the entirety of the text for an article or book is not indexed; instead, **the database indexes each document's record**, which includes things such as the article title, journal title (if appropriate), author, subject heading, and topic. Therefore, search terms become very important and should be thought of as words and phrases that you expect to appear as part of one of the record sections (title, topic, or subject heading, for example). Attempting to search for "it was the best of times, it was the worst of times" in a catalog or database might return an article entitled "It was the best of times, it was the worst of times" as its most relevant record, while a search engine, which has indexed the entirety of each web page will return pages that have information describing the phrase as being the beginning of the famous opening sentence from Charles Dickens's *A Tale of Two Cities*.

Another big difference in using catalogs or databases versus search engines is the relevance and currency of your results. **Catalogs and databases are maintained by information professionals**; therefore, the trustworthiness of the resource has been verified. Searches can be tailored to return results from specific publication periods, ensuring that the information you are finding is appropriately current. Although searching by date is also possible through search engines, they do not always filter explicitly by publication date. Furthermore, catalog and database results do not point to a specific website, which may change or simply disappear depending on how long it has been since the site was last indexed; they point to a record, which can subsequently be used to locate and secure the published document. While it is still up to you as the researcher to determine the relevance of the resource to your specific research topic, some of the previous concerns we had with search engines are alleviated by having these curated lists.

Returning to our solar roadways example, let's focus on the 1,980 matching items we found for our query from our university catalog. Remember, we narrowed our results to this number by focusing only on articles from scholarly publications. As much as this is an improvement from 1.13 million results, it is still a lot of material to look through. Certainly, using additional search tools can further refine our query by perhaps limiting where our query terms appear in the record or searching for documents published in 2013 or later, but what if we only want results that specifically pertain to engineering and not other academic disciplines? To do this, we can use journal databases catering to the specific discipline in which we are interested.

Engineering in Action

Enter the search term "history solar roadways" (without the quotes) into a database available through your university's library website that is specific to engineering, and look at the generated results. You may need to talk to your university librarian if you are unsure as to how to access these resources. How do these search results compare to the results you generated using either a general or topic-specific search engine? How does it compare to the results you generated from your University's library catalog?

Library databases are much like library catalogs but are not specific to the holdings at a given institution. They contain record information from published works, just like we saw for the library catalog, but are often designed to be subject-specific, such as an engineering-specific database. Just like the catalog search, we are often not interested in searching for specific phrases but using specific search terms we expect to find in the database record. For the purposes of illustrating this, we will be using Engineering Village from Elsevier, which focuses on "engineering literature from reliable sources, including journals, conference proceedings, trade publications, patents, government reports, and articles in press."[9] Searching for "history of solar roadways" in Engineering Village results in 0 records found. Part of this is due to the databases Engineering Village uses. As we continue to narrow the scope across which our search term is being used, fewer and fewer results may show up. In the case of this particular search, no documents were found within the databases searched that contained those terms in their record fields. Removing the term history from the search query resulted in 159 records, as shown in Figure 4.3. As you can see from the figure, the documents within the first page of results included conference and journal articles. In other words, the nature of the search database resulted entirely in scholarly documents. Database searches may suffer from a similar issue to what we saw in our search engine results in that you may not have access to all the documents at your specific university. How you obtain documents that are not owned by your university library is beyond the scope of this text, but we encourage you to engage your university librarians to see what options you have for obtaining scholarly work that is not immediately accessible by your institution, as options exist that may reduce or eliminate the cost of access to you as a student.

So, Which Search Should I Use?

As with most things in life, there is no one right answer to which search you should use. Our honest opinion is that it depends. Generic search engines might be fine for beginning your search to see what is out there on the internet currently and may be an excellent source for

---

[9] https://blog.engineeringvillage.com/about/.

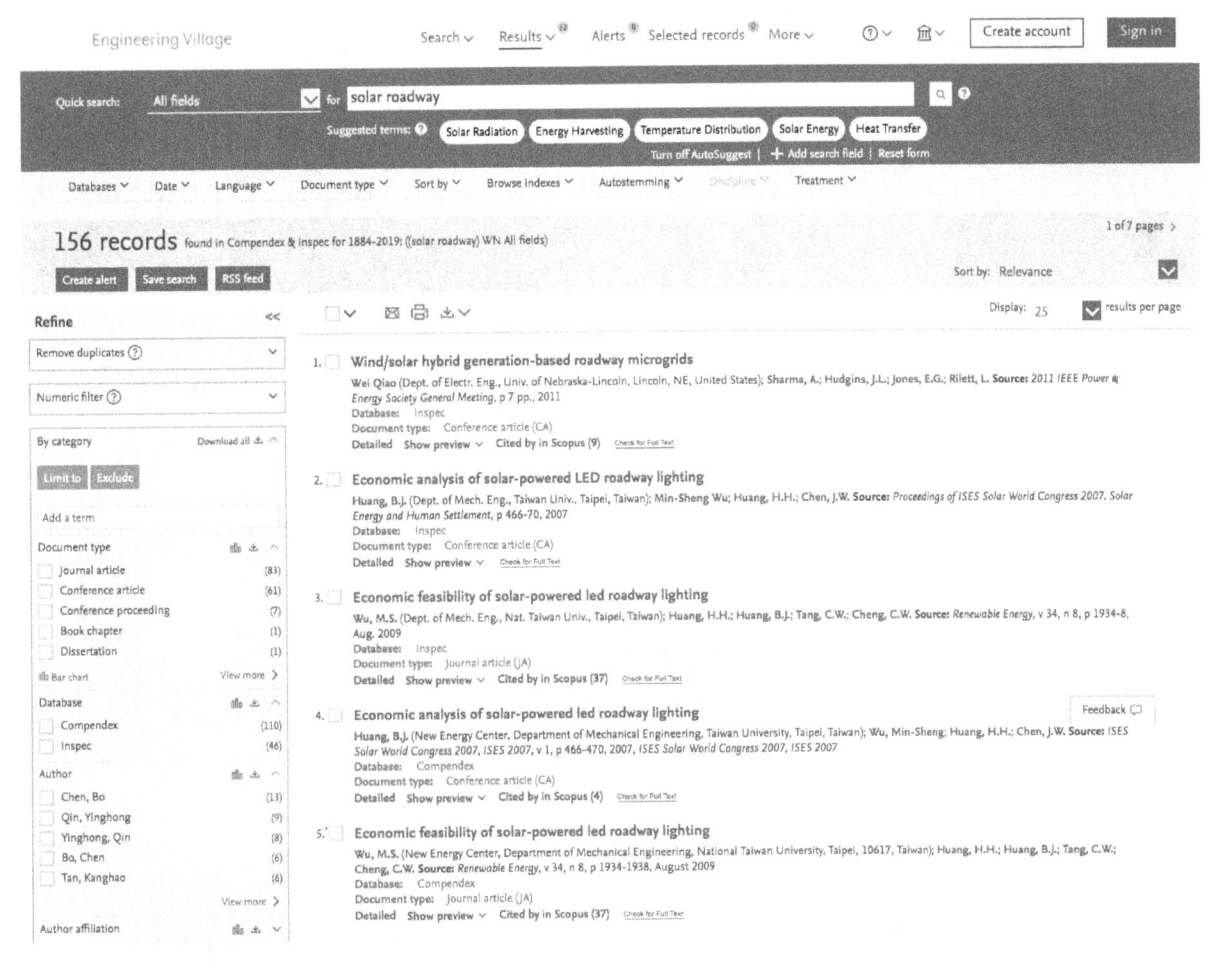

FIGURE 4.3 Results from library database (Engineering Village) search for "solar roadways"

information that has not been published. Searches using Google Scholar or Google Patents narrow the websites from which the search engine pulls information quite drastically, which can be helpful once you have refined your search and are looking for specific resources. University library catalogs and databases offer a curated set of results, guaranteeing their authenticity and pertinence. While catalogs contain documents that are accessible to university students, not all the digital resources found in databases will be automatically accessible; as such, you will need to work with your library to determine what means you as a student have in obtaining these resources. For your convenience, we have included Table 4.1 in an effort to summarize this information.

TABLE 4.1 Advantages and Disadvantages of Available Search Resources

Search Resource	Advantages	Disadvantages	Scholarly?
Search Engines/Personal Assistants (Google, Bing, Alexa, Yahoo)	• Breadth of search • Includes unpublished materials • Advanced search options can provide search specificity	• Results based somewhat on popularity; material not vetted • Links might be outdated or behind paywalls	Maybe—results might include scholarly work but not by default
Specialized Search Engines (Google Scholar, Semantic Scholar)	• Results limited to education and scholarly websites • Results focus more on documents versus websites • Includes unpublished materials	• Results are not curated • May be unclear as to the nature of the linked document (peer-reviewed publication versus non-peer-reviewed) • Linked documents may not be accessible to your institution	Somewhat—results come from scholarly-like websites, may or may not be peer-reviewed
Library Catalog and Databases	• Search specifically focuses on published materials, thus improving authenticity of materials listed • Curation of results can provide more relevant search results than search engine algorithm • Database specificity allows for carefully crafted search results	• Databases may not be all-encompassing, requiring more than one search • Not all linked documents may be accessible to your institution • Unpublished materials (conference posters, etc.) are not likely to show up in these systems	Yes—Lists are curated, and scholarly material can easily be identified

Over the course of your time in engineering, you will likely become fluent in using one or more of these tools as you conduct research queries. You will find what works best for you and will begin to realize that all of these tools have their pros and cons. Ultimately, making sure you can find and appropriately use the most pertinent information is one of the biggest lifelong learning skills you will acquire in the course of your academic career and will likely be one of the top skills you utilize throughout the course of your professional career.

READING SCHOLARLY ARTICLES

Being able to effectively read technical articles actually begins when you are searching for them. Titles are often the first part of an article or website that catches your eye. In certain situations, these are specifically designed to capture your attention (consider some of the more extreme newspaper headlines or the material typically included in a movie trailer). In a more academic setting, these titles trend towards being overly specific, including complex terms and, at times, undefined acronyms, making the potential value or relevance difficult to determine for anyone not well versed in a specific field. Whether it is attempting to decipher the sometimes overly complex titles or trying to decide if an article helps answer the question you are searching for, many times, the title itself is not enough. Rather than just looking at titles or introductory paragraphs, one of the best ways to search is to go through and read the abstract. While it may take you a little bit longer during the search to do this, you are more likely to get references that are closer to the topic that you are looking for than if you go by the article name alone. The abstract, if well written, is designed to tell you everything that is in the article. It is the best summary that you will find about that material and provides a more robust picture than that which is obtained from the title alone.

Best practices exist at every stage of research, including reading scholarly articles.

We mentioned in the introduction of this chapter the non-linear nature of information searches. While this is true, what is also true is that best practices exist at every stage of research, including reading scholarly articles. When learning something for the first time, there are always several different suggestions about the best way to go about doing something. Rather than giving you just a single best way, what we would like to do is propose several different ways that you could go about reading technical articles. Based on our research as well as feedback from our students, there really is no one single way that works for everyone, so trying different methods can be a great opportunity to find an approach that works best for you.

Methods for approaching technical articles include:

- variations in how many times an article should be read,
- the order in which sections of the articles are read, and
- even the specific focus of the reading (i.e., what you should be thinking about when you are reading the article).

One of the most overwhelmingly supported suggestions that we hear from students when reading technical articles (often given in a tone of surprise) is that you really need to read the article more than once. This is not a book you might read for enjoyment, an article from a magazine that you might pick up, or something that comes across Twitter or BuzzFeed that you can read in a casual manner. It is something that requires a really clear focus. Often what is needed is a low distraction environment that results in fewer disturbances than what you might need when you are reading casually and multiple reads to truly digest and comprehend everything that is being said in the article. **Do not be surprised if it takes you three or even four times reading through an article to appreciate the full value of what is being presented.** Accomplishing multiple readings takes time, however, and it is often more valuable if these readings are spaced out (i.e., do not finish the article and immediately start back at the beginning) and are combined with a different approach for each reading.

Some people will suggest that these readings need to be full readings (all-the-way-through every time), others will suggest skimming the first time and then reading in more detail, and still others will suggest reading in more detail and circling or highlighting sections, somehow denoting key areas so when you go back and skim the next time you can really focus on those points rather than the entire document. Depending on your experience and knowledge of the topic, you may be able to skim over methodology sections (particularly if they mirror the approach from other sources) and instead focus on how the results and conclusions differ from a previous article. In contrast, maybe it is the new methodology that differentiates the paper from others you have read, and thus it should be read more closely. Some suggest reading the introduction and then focusing on the figures, while others find skipping the figures on the first reading and focusing initially on the text to be more logical. If you are less familiar with the work, the goal of the first reading may be to identify words and concepts with which you are unfamiliar so that time can be spent defining these terms or gaining additional knowledge from other sources on the concept before completing a second reading of the article.

Engineering in Action

Discuss with others approaches you have used to comprehend technical information. What methods worked best for you? What did you not find helpful? Did the effectiveness depend on the topic (or your interest in the topic)? Is there a method that someone mentioned that you might like to try?

Ultimately, it is important to recognize that reading technical articles takes time. Even if you are an avid reader, the person who can devour a Harry Potter novel in an afternoon, technical articles are not typically read at the same speed as books read for pleasure. I have been doing this for more than twenty years, and I can still see a demonstrable difference between how long it takes me to read one thousand words of a New York Times Best Seller (about two minutes) and one thousand words of a technical article (closer to five minutes).

SYNTHESIZING INFORMATION

To *synthesize information* is to tie it together in a new way, make sense of information from multiple sources, and combine it together to support or refute the point we are trying to make.

Once we have found multiple sources we want for our topic and we've read through all of them, our next step is to actually start the writing process. When we talk about *synthesizing information* from articles, this is more than simply pulling out a quote that we like or rephrasing some of the information we found. Synthesis requires that we truly understand and comprehend what each of these articles is talking about prior to using them. To synthesize information is to tie it together in a new way, make sense of information from multiple sources, and combine it together to support or refute the point we are trying to make. This means that we cannot just have a basic understanding of what is going on, and we cannot have large gaps in our understanding of the portion of the sources we intend to use. While it is acceptable, and even encouraged in some cases, to use direct quotes when writing a review or summary article, ultimately we need to make sure that the quote is saying exactly what we want and being included in a meaningful, logical way. If we try to force our sentences to fit a quote that may not really need to be directly quoted but instead can be summarized and still referenced appropriately, we often end up with paragraphs that sound stilted, awkward, and have a weak overall flow. If we are referencing material we do not fully comprehend, our writing often reads as disjointed, confusing, or (in some extremes) factually incorrect.

Many times, the best way to synthesize information is to start by asking what story we want to tell. We can then decide what points we are trying to make and establish an outline that guides the path our story will take. Once this general framework exists, we can then look back to our sources to see how they support our key points. For example, are there multiple articles that support a common point, pieces of information that add slightly different knowledge, or works that may contrast what is being said?

Once we know what it is we want to say and we have gone through our articles and annotated (e.g., highlighted, starred, underlined) specific points, whether a direct quote, a figure, a graph, or a specific piece of data that we really want to use, we can then start making connections between the information that we want to include and the story we want to tell. Sometimes it seems counterintuitive to think of a technical article as a story, but for most of us, stories are something that we enjoy reading and enjoy hearing. To approach a technical article in this manner and yet still maintain a professional tone and tenor (which we will talk about in greater detail in Chapter 5) allows or improves the overall flow of our article. This is incredibly critical when we are trying to capture the key points we want to highlight and to remain focused on justifying the point we would like to make.

PLAGIARISM

As we begin to fill in our outline with supporting evidence from our sources, we need to constantly be aware of our ethical responsibility to avoid plagiarism. *Plagiarism* is probably a word that you have encountered in the past, and when you hear it, the first thing that likely comes to mind is some of the more extreme forms of plagiarism: someone claiming credit for something they did not write or including a long section of text in a document without acknowledging the source(s). However, plagiarism is a bit more nuanced than that, and this section will talk about those different levels of plagiarism. Defined as "an act or instance of steal[ing] and pass[ing] off the ideas or words of another as one's own" by Merriam-Webster,[10] plagiarism is often included in a college or university's code of conduct. Many times, that code of conduct provides specific definitions of plagiarism that are most often encountered on a college or university campus. If you have never had a chance to look over your own college or university's code, we encourage you to locate it either on your campus website or from a campus resource and recognize the potentially serious implications that can result if you intentionally or unintentionally plagiarize (or break any of the expectations laid out in that code of conduct for that matter). For example, at our university, plagiarism can result in something as mild as a warning or a zero on that particular assignment to more severe penalties such as failing the course in which the assignment was submitted or even possible suspension or expulsion from the university in extreme cases.

> **Engineering in Action**
>
> Locate your college or university's code of conduct. This can often be found online, and many times a reference will be included in your syllabus. Some schools require students to sign an honor code. Even without this level of formality, as a student you are bound to these expectations. Take a look at the definition of plagiarism and some of the potential consequences that may be imposed on those violating these expectations. (Note: It's not a bad idea to look over the entire document. You may be surprised at some of the expectations.)

Since plagiarism is such a serious issue, let us take some time to distinguish between the different levels of severity. Figure 4.4 categorizes five levels of plagiarism, from the most severe at the top to the least severe at the bottom. Since individuals often recognize plagiarism in its most severe form first, we will start with the highest or most obvious form of plagiarism and work down to those instances which, from a technical definition, are still considered plagiarism but are often unintentional or unclear. Our goal with this is to make you more aware that even

---

10 https://www.merriam-webster.com/.

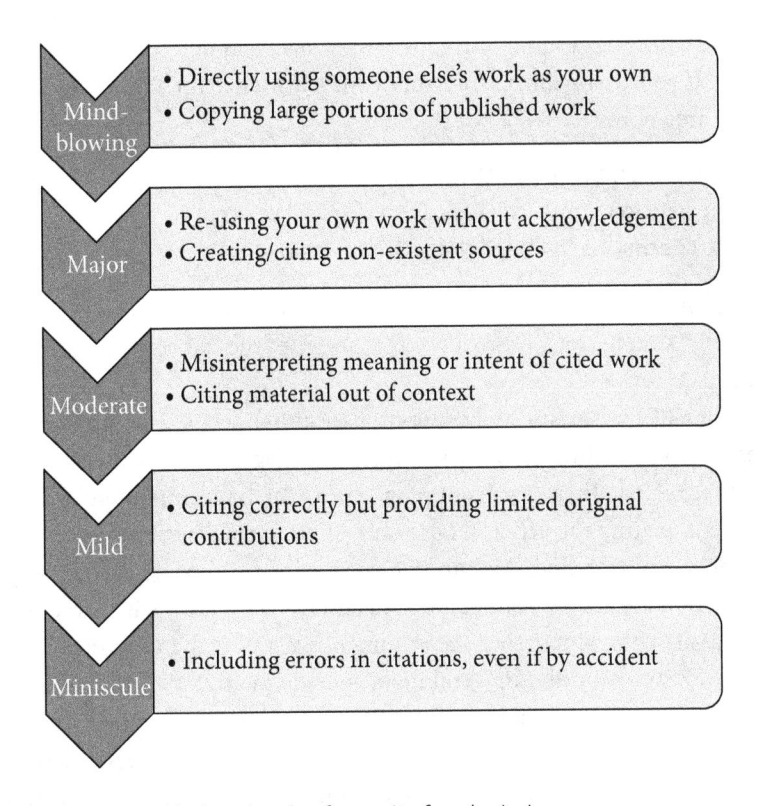

- **Mind-blowing**
  - Directly using someone else's work as your own
  - Copying large portions of published work
- **Major**
  - Re-using your own work without acknowledgement
  - Creating/citing non-existent sources
- **Moderate**
  - Misinterpreting meaning or intent of cited work
  - Citing material out of context
- **Mild**
  - Citing correctly but providing limited original contributions
- **Miniscule**
  - Including errors in citations, even if by accident

FIGURE 4.4 Various levels of severity for plagiarism

these "lighter" instances still classify and qualify as plagiarism, and as such, you need to exercise a more cautious approach as you develop your own individual work.

The mind-blowing level of plagiarism follows the definition in its purest form: the material you are saying is your own was developed by someone else. This is true not only if it is pulled from published sources but also if it was either created by someone else and copied by you (e.g., copying a homework solution from a friend) or created at your request by someone else and labeled as yours (e.g., paying someone to write a paper for you).

The second, or major level, of plagiarism, addresses the often-asked question of "can I plagiarize myself?". The response to that question is yes. On first consideration, it seems a little strange to think this is possible, since the very definition states "the ideas or words of **another**," and if they were my words to begin with, then no "other" person is involved. Once those words have been written and published or submitted for evaluation (e.g., turned in for a grade), using them again for a different purpose (i.e., another paper or assignment) means they are no longer original for that second purpose. It is this concept of originality that dictates the plagiarism classification. This situation is avoided by approaching your own material as you would others, appropriately citing the previous work and recognizing it as a portion of the submission that is cited rather than original. The second example under the major level is much more straightforward: do not make up sources. If you are unable to find supporting evidence for a portion of your submission, consider one of three points:

1. Does this point require supporting evidence,
2. Is it appropriate to express the idea as an original concept or opinion, or
3. Should the concept be included in the submission at all?

If the idea is one that needs justification and should not be expressed as an opinion, a reputable and valid source must be included; otherwise, the concept should be removed from the submission.

Moderate forms of plagiarism occur most often either when a complete understanding of the source has not been achieved or when the limitations of the source information are not recognized and respected. These forms of plagiarism may be unintended and often result from a lack of time spent reading the articles. Less time often means fewer or faster reads through the article, which can result in reduced comprehension of select aspects. Sometimes, rather than a complete reading of an article, what occurs instead is a brief skim, with a focus on locating key words that may assist with supporting the argument we are trying to make. Consider an article that states that over 80% of individuals taking a specific medication in the study lost weight. If we were skimming the article to determine the impact of the medication on weight, we might think this is a valuable point to include. But consider if, earlier in the same section, the article commented on the high levels of nausea and vomiting experienced by virtually every participant in the study. Our cursory look may cause us to attribute the weight loss to the medication directly, rather than as a side effect of the nausea patients experienced from the medication. Knowing and appreciating the limitations of information is also valuable. Consider this statement: I am often the tallest person in the room. From this statement alone, one may be tempted to conclude that the individual is above average in height. But what if the person is barely five feet tall and a kindergarten teacher? In this case, the statement is reasonable when taken in the context of their classroom but likely less authentic if the same individual is placed in a room of their peers where, in the United States, the average height for adult women is 5'4" and for men is 5'9".[11] While these examples may feel a bit obvious, the intent is to demonstrate the ease with which these moderate forms of plagiarism may develop.

Plagiarism in the mild form may not even really look or feel like plagiarism at first. After all, the paper includes multiple citations, correctly attributes work to the appropriate author(s), and the source information is correctly interpreted and taken in context. Upon closer inspection, however, what is found is that virtually every key thought or idea fundamentally is attributed to someone else. In many ways, this form of plagiarism goes back to the originality concept presented previously. **The work you are creating should provide some level of original**

---

11 https://www.cdc.gov/nchs/fastats/body-measurements.htm.

contribution to the topic under discussion to be considered an authentic work. Otherwise, what you are creating is just a mash-up of the work of others (similar to the Beca approach in *Pitch Perfect*—interesting, but not what impressed Keegan-Michael Key's character).

The lowest level of plagiarism (minuscule) is often the most unintended form. In cases such as these, we have integrated information from a high-quality source into our predominantly original work, included quotation marks if necessary, an in-line reference where appropriate, and a reference within our bibliography (or reference section) at the end of our work. What is missing, however, is the absolutely correct form of the reference in the format required for the submission. Maybe we transposed the page numbers, forgot to include the issue number, shortened the title, or misspelled an author's name. Depending on the severity of the error, the result could be the inability of others to locate the reference using the information provided. To a lesser extent, the inaccuracies can potentially delay finding the original source. In either instance, it suggests a lack of professionalism in the creation of your document.

REFERENCE MANAGERS AND CITATIONS

Throughout this chapter, we have outlined things for you to consider when searching for information to better understand and perhaps even report on a topic of interest. This could be something you are working on as part of a course project or homework, for research, or simply just for your own knowledge. As you might imagine, particularly for larger projects or times when more in-depth knowledge is required, the scope and number of resources you might use to obtain your knowledge could be significant. If you look at a review article in a scientific journal, it is not uncommon to see the author(s) of those articles cite upwards of fifty references or more. Using this many references not only gives credit to the original work that was used to support the review article, but it also reinforces that the arguments made within the review article are supported by data and facts discovered by independent, validated studies. As you might imagine, keeping track of more than fifty references, including where each document is located (for digital documents, is it on a cloud drive, flash drive, your computer's hard drive, etc.), what information can be found in each document, and then appropriately citing them in your paper in a specific format can be a monumental task. Although one can do all of this manually, there are computer programs and web applications that are specifically designed to help you with this part of your research. As we will discuss in greater detail below, reference managers not only help you organize and manage your references but also make it easier to cite those works within your document. Although we will only describe one manager in detail, we recommend talking with your university librarian to find out what resources/products are available to you at your home institution.

Reference Managers—What Are They?

As the name implies, *reference managers* do exactly that; that is, they help manage the resources you have compiled in support of a project, research topic, and so on. As you acquire information about your topic of interest, these programs can help you keep those references organized and compiled in a centralized location. Depending on the manager you choose to use, this centralized location might be somewhere in the cloud or on your local computer. Other differences among reference managers that you might consider include what operating system and word processing programs you and your team use. Table 4.2 lists some (not meant to be comprehensive) reference managers the authors are aware of or have used, as well as some of their features. Within the table itself, integration refers to the word processing applications the reference manager works with, while team share refers to whether or not libraries (a collection of reference materials) can be shared among multiple people. As you can see, most modern reference managers allow for storage of PDF documents and sharing of libraries among team members, although the exact mechanisms inherent in both these features will differ from program to program.

TABLE 4.2 Reference Managers and Their Features

Reference Manager	Cost	Integration	Compatibility	Web/Local	PDF Storage	Team Share?
EndNote	Commercial	MS Word, OpenOffice, LibreOffice	Mac, Windows	Both	Yes	Yes
Mendeley	Free	MS Word, LibreOffice	Mac, Windows, Linux	Both	Yes	Yes
Paperpile	Commercial	Google Docs, MS Word	Google Chrome, iOS, Android	Web	Yes	Yes
ReadCube/Papers	Commercial	MS Word, Google Docs	Mac, Windows	Web	Yes	Yes
RefWorks	Institutional	MS Word, Google Docs	Mac, Windows	Web	Yes	Yes
Zotero	Free	MS Word, LibreOffice, Google Docs	Mac, Windows, Linux, iOS	Both	Yes	Yes

Engineering in Action

Does your college or university suggest or encourage the use of a particular reference manager? What support do they provide? (Note: A good starting point for this search is your university library).

Using a Reference Manager

The first step in using a reference manager is importing references to manage. As shown in Figure 4.5, most programs will allow you to do this by uploading or including a PDF, uploading or including a formatted bibliography list, or manually creating a reference. Many searches we have described in this chapter can output the results of your search as a formatted bibliography list, which can then be uploaded to the reference manager. Alternatively, if you are planning on using your reference manager as a means of organizing your collection of PDF files, uploading or including a PDF might be the best way to go.

For most programs, metadata from the imported PDF can then be used to create a full reference for the document (Figure 4.6). In addition to organizing your PDFs and keeping them in a centralized location, most managers will also allow you to read and annotate (highlight, note, etc.) your PDFs directly within the program. Depending on the reference manager used, this information can even be shared among collaborators, providing multiple users and devices access to a singular location for references and annotations.

Once you have read and compiled your references, you can also use your reference manager to help you with writing your paper by keeping track of your in-line citations. As we will see in the next section, in-line citations are an important part of your documentation practices (think back to the plagiarism discussion), and reference managers are a tool in your toolbox to ensure you are appropriately citing the works of others in your own writings.

Citations

As we describe in the section regarding plagiarism, citations are used to attribute non-original information to the appropriate author(s) (which might include yourself). This practice not only acknowledges the original source and provides the reader a means by which to find the original work but also lends credibility to your own argument. In other words, **your own ideas and arguments are strengthened by the fact that other independent researchers have drawn similar conclusions or have generated data that supports or corroborates your position.** With this in mind, having a broad and deep understanding of the literature and data already available is a crucial part of the research process, and appropriately synthesizing this information can directly impact the overall technical quality of your research paper or report.

To appropriately cite a work in the creation of a document, you must include two things: **a reference at the end of the paper and an in-line citation indicating where the reference was used.** The style of the in-line citation can vary depending on the requirements of the document or instructor, so it is always advised that you determine what style is required prior to starting your paper. Although some word processors have a bibliography tool, we find that many students choose to insert in-line citations and add references to their document manually. This may work well for shorter documents that are created individually, but for longer, more technical documents, or for documents that were created or worked on collaboratively, this method can often be difficult to manage, particularly if the in-text citation style uses sequential numbering. If

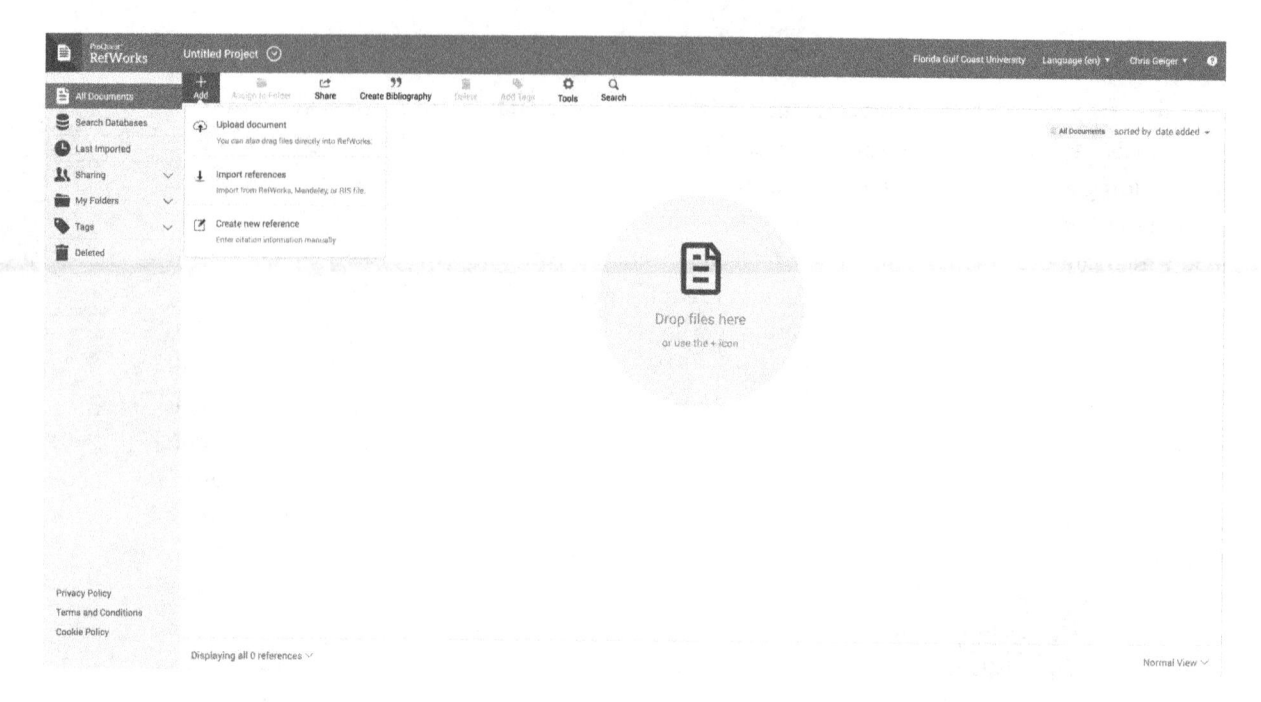

FIGURE 4.5 Importing references into a reference manager

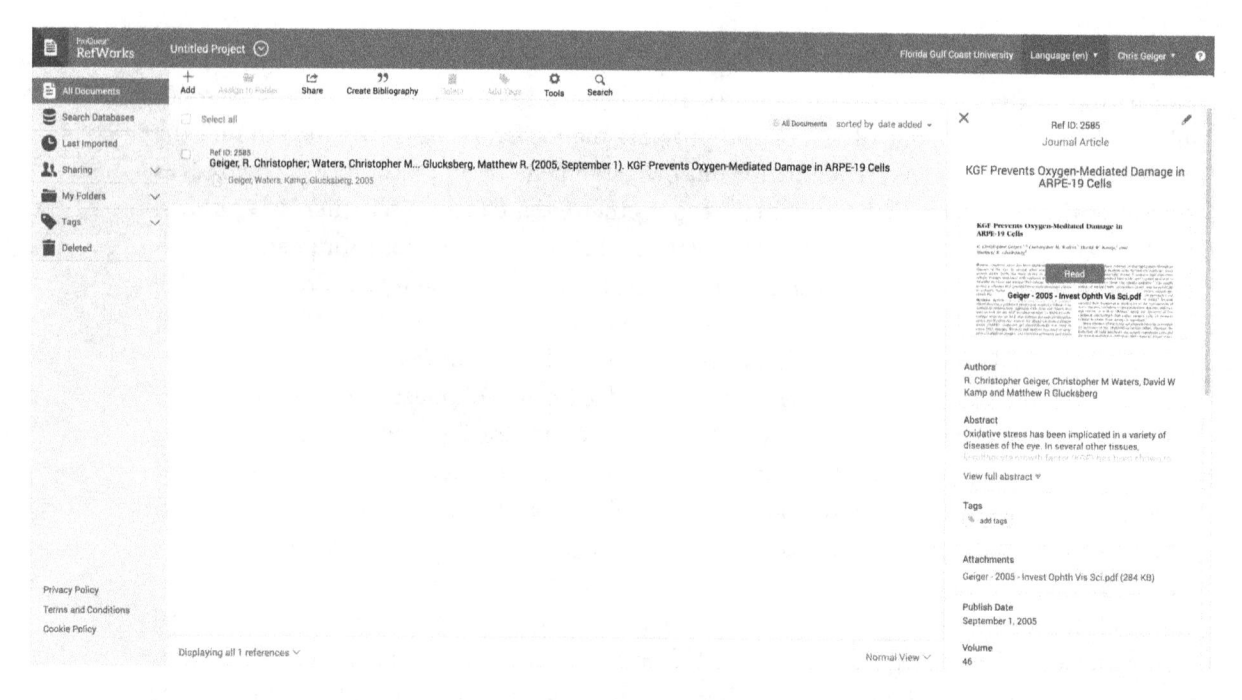

FIGURE 4.6 Using PDF metadata to generate reference information

104   FOUNDATIONS OF ENGINEERING

you are already using a reference manager to help organize your references and annotations, you can also use it to help with the citation process by directly citing a reference from your reference manager library within your word processing application. Since you can use results from your library searches or information scraped from imported PDFs to create each reference, you will not need to manually enter the reference information for each document, saving you time and potentially minimizing mistakes. Furthermore, the reference manager will keep track of all your references, making sure they are appropriately ordered and formatted within the document. Although the exact mechanics of this process varies from program to program, there are many tutorials available online and likely resources available on your campus that can help you with any questions you might have.

CHAPTER SUMMARY

This chapter focuses on *information literacy*: knowing when information is needed to solve a problem or make a decision, knowing where to find that information, and ultimately how to use it. Within this chapter, the following topics were covered:

- The pros and cons associated with electronic resources that are likely available to you as you go in search of technical information as well as how to best utilize those resources to ensure the information you find is authentic and appropriate
- Recognizing *peer-reviewed material* and how this work (which is reviewed by external content experts) often has a higher level of validity than work that has not been peer-reviewed
- Helpful hints regarding how to read scholarly or technical materials and things to consider as you begin to *synthesize information* from multiple sources
- Understanding the concept of *plagiarism*, including how it is possible for you to plagiarize yourself, and how to appropriately use citations to acknowledge the work done by other authors
- How *reference managers* can help you organize your supporting documents or materials for a project or paper and help with the citation process

END OF CHAPTER ACTIVITIES

Individual Activities

1. Conduct a search on your major using the following variations:
    a. Simply type your major without quotes (e.g., Mechanical Engineering), then use various combinations of operators (e.g., AND, OR, NOT) into Google or your search engine of choice. How do the results vary among these searches?

b. Search your major in Google Scholar and compare these results to those you received in part a.
   c. What happens if you enter your major as a search term in your library's catalog?
2. Identify three to five databases in your library that target engineering searches. What are the primary similarities and differences between these databases?
3. Search for "Panama Canal" in three of the databases you identified in question 2.
   a. What sources are identified in multiple databases?
   b. Which database provides the greatest variety in terms of *type* of source (i.e., book, journal, movie, etc.)?
   c. Which database provides the largest number of sources?
   d. Which database provides the most current sources?
4. Select a journal article from your Panama Canal search and read it. What approaches discussed in this chapter did you find most effective when reading this article?
5. What options do you have if your library does not have the book, article, resources that you need? What is the process to request an item, and how long does it typically take to receive it?
6. Write a brief (~300 word) response to this question: Why should I choose to go to school at your university? Your response should be based on facts (not opinions) and contain at least three appropriately cited references.

Team Activities

1. Swap your response from question 6 in the individual activities with a team member. Investigate whether the response contains plagiarism. If you think it might, what level would it be? What adjustments could be made to eliminate plagiarism from the document?
2. Have all team members share their responses to question 6 (from the individual activities above). As a team, combine these to a single (~500 word) team response.
3. Have each team member present three to five main ideas from the Panama Canal articles you read from question 4 in the individual activities. Create an outline for a paper that combines the ideas from all group members.
4. As a team, conduct a database search on sustainable development and select a journal article from your results. Have the entire team read the article and, as a team, create a 500–750-word summary of the key points in the article.

CHAPTER REFERENCES

Appendix A—Information Literacy Competency Standards for Higher Education—The Association of College and Research Libraries (ACRL). 2005. In *Information Literacy: A Practitioner's Guide*, 143–52. Chandos Publishing. http://dx.doi.org/10.1016/b978-1-84334-065-2.50016-3.

Presidential Committee on Information Literacy: Final Report. 1990. *American Library Association*. http://www.ala.org/acrl/oublications/whitepapers/presidential.

Rhodes, Terrel L. 2010. *Assessing Outcomes and Improving Achievement: Tips and Tools for Using Rubrics,* 51. Association of American Colleges.

Schwartz, Barry. 2016. "Google's Search Knows about over 130 Trillion Pages." *Search Engine Land.* Accessed February 26, 2019. https://searchengineland.com/googles-search-indexes-hits-130-trillion-pages-documents-263378.

Weiner, Sharon A., and Lana W. Jackman. 2010. "The National Forum on Information Literacy, Inc." *College & Undergraduate Libraries* 17, no. 1 (March): 114–20. http://infolit.org.

CREDITS

Fig. 4.1: Adapted from: Association of College & Research Libraries, "Framework for Information Literacy for Higher Education," http://www.ala.org/acrl/standards/ilframework. Copyright © 2015 by Association of College & Research Libraries.

Fig. 4.2: Copyright © 2019 by Florida Gulf Coast University.

Fig. 4.3: Copyright © 2019 by Elsevier.

Fig. 4.5: Copyright © 2019 by ProQuest LLC.

Fig. 4.6: Copyright © 2019 by ProQuest LLC.

CHAPTER 5
WRITTEN COMMUNICATION

BEFORE YOU BEGIN

Take a minute to consider the following questions:

- What are some of the characteristics of good writing?
- How does professional/technical writing differ from creative writing?
- What experience(s) have you had with professional written communication?
- How can an outline assist in the development of a written document?
- What do you think of when you read the phrase "significantly different"?

LEARNING OBJECTIVES

By the end of this chapter, you should be able to:

- Identify key characteristics of effective professional writing
- Improve future written communication by implementing a final editing process
- Recognize the level of professionalism required for various types of written communication and apply these levels correctly to given situations
- Integrate audience expectations into outlines used as the starting point for developing written communications
- Incorporate variables, abbreviations, figures, tables, and equations into written communications in an appropriate manner
- Express numerical values in a readable format that employs the correct significant figures
- Differentiate between general differences and statistical differences and adjust writing to clarify this point
- Implement appendices in an appropriate manner to increase the clarity and organization of written communications

INTRODUCTION

Many of you reading this book may have already taken one or more courses in writing or composition as part of your general education requirements early in your college career. Better yet, some of you may already have had a course in technical or professional writing or may be scheduled to take such a course in the near future. This chapter (and the next one, which will discuss oral communication) is not designed to replace those types of courses; however, we do hope that we either reinforce the lessons learned in those other courses or provide an extension to or perspective on their material that you find useful. Our scope for this chapter is written communication as it pertains to engineering, and while our focus will be primarily on the types of communication you will encounter while completing your education, the framework we will present can easily be translated to all professional communication post-graduation.

We like to think of writing much in the same way designers think about fashion: there is a personal nature to communication that will make what you write uniquely yours, but in the same sense, good writing (just like good fashion) follows certain guidelines, basic structure, and form. While stylistic opinions can vary, and as such, it is possible that some of our recommendations may run counter to faculty, clients or superiors within your own organizations; as we have suggested in previous chapters, these suggestions are based on our experiences as faculty members and are the same recommendations we make to our own students at our home institution. In the end, as a writer, it is your responsibility to understand how best to implement these recommendations within your own sense of style while being conscious of your audience and their communication requirements.

While the way you write will make what you write uniquely yours, good writing still follows certain guidelines, basic structure, and form.

WRITING CLARITY

One of the biggest mistakes students make when it comes to technical or professional communication is that their writing often lacks clarity. This lack of clarity can be due to a number of issues; however, we are going to focus on five of the issues we find most often in our own student's writing. In no particular order, those issues are improper sentence structure, being too verbose, not being specific enough, not thinking of a report or paper as a story, and not being consistent across the totality of the report (often a significant issue for group submissions). In the paragraphs that follow, we will look at each of these issues and offer some suggestions as to how to avoid them when writing technically.

Sentence Structure

One area where we see students struggle in technical communication is sentence structure. Historically, written communication for lab reports, journal articles, and any other type of communication discussing testing or research was expected to be written in the *passive voice* (although this is again one of those areas that is changing,[1] and you should discuss expectations with your instructor or TA). Take the following sentence, for example:

We measured out 2 g of sodium benzoate and added it to a 50 mL beaker.

In the passive voice, the sentence would be restructured:

Two grams of sodium benzoate was added to a 50 mL beaker.

When using *passive voice*, the object of the action becomes the subject of the sentence, rather than the actor. In our example, "We" (the actors) are the people that measured the sodium benzoate, and in the active voice sentence, "We" are in the subject position of the sentence. Using passive voice, the sodium benzoate (the object of the action) assumes the subject position of the sentence. While most students are familiar with active voice, many have learned to avoid passive voice, as it is typically seen as being a "weaker" sentence format. It likely does not help that the word *passive* is synonymous with weakness. However, we understand that passive sentences can either remove or obfuscate the actor (which is actually one of the reasons to use them in technical writing), and they limit sentence variety, which can decrease readability.

Technical writing has relied on passive voice because either the action or object being acted on is more important than the actor, or the identity of the actor is obvious (often the author or writer), and as such, it is not necessary to refer to them within the sentence. Passive voice is seen most often in the Materials and Methods section of a report or paper, where a) it is obvious who did the acting (the authors of the paper) and b) the sentence emphasis should be on the action versus the actor (i.e., it is more important to tell me that 2 g of sodium benzoate was added to a 50 mL beaker than telling me that "we" measured out that chemical and added it to the beaker). The passive sentence emphasizes the action taking place, and concisely provides the reader with the needed information.

---

1 David Banks, "The extent to which the passive voice is used in the scientific journal article, 1985–2015," *Functional Linguistics*.

> The passive sentence emphasizes the action taking place and concisely provides the reader with the needed information.

One difference in the example provided and examples you may have come across in previous writing courses is that in our case, we have eliminated the actor entirely in our sentence, as it was obvious that the individual(s) conducting the experiment was the person who added the sodium benzoate to the beaker. In many writing courses, instructors will emphasize that the passive voice is actually less concise than the active voice, primarily because the actor remains in the sentence. Examples are often given as follows:

Active: The Eagles scored a touchdown after taking their last timeout.

Passive: A touchdown was scored by the Eagles after taking their last timeout.

This is one of those sentences in which the actor is necessary to understand the meaning of the sentence. Without stating which team scored the touchdown, it would be impossible for you as a fan of the Eagles to understand if this should make you happy or sad. Therefore, writing in the active voice is going to be the most concise approach and is obviously easier to read. In technical writing, when these situations come up, it is best to talk to your supervisor to understand their expectations, but we would say this is a scenario where using the active voice makes sense.

One additional area where students struggle with sentence structure in using the passive voice is with respect to tense. **For the most part, reports are primarily going to include present tense when describing concepts, theories, and statements of fact and past tense when describing things that have already happened, such as protocols and results.** Note in our example that we stated that the salt *was* added to the beaker. Since we had already completed the experiment (otherwise, how could we have results), those sections describing what was done should be written in the past tense. Making sure this is maintained across the report, and particularly within a given paragraph or sentence, should be something to keep in mind and routinely check for during the editing process (more on that later).

Conciseness

Of all the common fallacies we see in report writing, the most common one is that students equate report length with report quality. There seems to be a false equivalency between good technical writing, report length, and the type of language used (we consider this the "more is better" fallacy). Being clear and concise (i.e., to the point) is something that will serve you well across all forms of communication throughout your engineering career. This does not mean that your writing should lack description or detail; these two ideas are not mutually exclusive. It does, however, mean that details should be limited to only that which is important to your report, and extraneous information should be eliminated. One way to test this premise is to critically examine your writing and see if it appropriately answers the simple question "why?". Remember to do this in the absence of prior knowledge. In other words, try to evaluate your report from the perspective of someone who has not completed the research and/or experiments you have completed. For example, if your report does not adequately express the reason a specific test or statistical analysis was run, or it does not provide enough details to support the conclusions you have drawn, it is likely that, in an effort to be concise, you have left out pertinent information that is necessary for the reader to fully understand the outcomes of your report.

Being clear and concise is something that will serve you well across all forms of communication throughout your engineering career.

Unless there is a specified word count or page number required by the format of the report or your supervisor (instructor), reports should only be long enough to completely and thoroughly describe the methodology, results, and conclusions of the activity in question. Again, details should be sufficient to appropriately describe the task or activity being completed; however, care should be taken to ensure that only information that drives the narrative forward in a meaningful way is included (see the next section, which describes treating your report as a story for additional information). Information that is tangential to the crux of the report should be limited or eliminated altogether. For example, if you are tasked with reporting the wet and dry weights of a sample to determine how much fluid it contained, describing how the tasks were divided amongst the group, what day the data was collected, or how additional tests were required to be conducted due to human error may not enhance the report, even though these activities might have taken up a significant percentage of the overall time you spent on the laboratory exercise. Much like a journal article has a clearly defined objective, and everything within that article supports the dissemination of that objective, so should other types of technical writing, unless you are explicitly told otherwise.

In addition to avoiding descriptions of activities that are tangential to the primary purpose of your writing, writing in a technical fashion does not automatically require the use of complicated polysyllabic words. Although technical content can be complicated and may include words that you do not come across in everyday conversation, and although there is an expected level of professionalism that should appear in your writing, these two things do not mean that your report requires the use of complex language to be effective. On the contrary, many reports will likely need to be written in such a way that a non-expert can understand your writing just as well as an expert in your field. Many times, we find these big words are coupled into long sentences, which do nothing but provide the reader a tortuous path in trying to understand the meaning of your writing. By shortening your language and writing simplistically, your readers will be able to understand the intent of your writing and will be able to understand and recall your conclusions more easily. Listed in Table 5.1 are six examples (three each) of terms and phrases we commonly see in report writing that can easily be shortened to enhance your document's readability. Remember, the goal of your writing should be to *express* your findings, not *impress* your reader or supervisor with your knowledge.[2]

TABLE 5.1 Examples of Words and Phrases That Can Be Altered to Enhance the Readability of Your Writing

	Lengthy	Shortened
Words	Ascertain	Learn
	Utilize	Use
	Terminate	End
Phrases	Due to the fact that	Because
	Prior to the start of	Before
	In the event that	If

Along these lines, we often find that students use the thesaurus to find alternatives to select words. While this approach can be beneficial for reducing redundancy in your writing (four sentences in a row that use the same verb does become repetitious to read), make sure that you fully understand any word you select as a replacement. Sometimes the thesaurus includes words that are similar but not completely the same, making it so all choices are not interchangeable in every context. Using a word that you are not familiar with in a context for which it is not appropriate becomes more confusing for your reader than using the same word multiple times. This goes back to ensuring clarity in your writing.

Precision

As important as it is to be concise in your writing, it is just as important to write precisely, particularly when you are trying to convey highly technical information. For our purposes, we will describe precision in the areas of quantification, specificity, and accuracy. In all of these cases, using imprecise language results in either requiring the reader to draw their own conclusions or perhaps even allowing them to misinterpret your results or statements.

As the name implies, quantification refers to using descriptors that provide as precise a value as possible in whatever it is you're describing. Words such as "few," "many," or "some" do not offer the reader a quantifiable number such as "three" or "ten." This level of precision is obviously

---

[2] Robert W. Bly, "Avoid these technical writing mistakes," *Chemical Engineering Progress*.

necessary when writing instructional or methodological documentation, particularly when quantities are important, but it is also extremely important across all types of technical writing.

Imprecise: Only a **few** patients had an adverse reaction to the medication.

Precise: **Twenty** patients out of the sample cohort of 1000 (2%) had an adverse reaction to the medication.

Recognize, though, that in many circumstances, quantification of a descriptor might not be appropriate for a variety of reasons. For example, in many cookbooks, you might find that they will tell you to allow your finished product to cool without giving you a quantifiable temperature as to what "cool" is. In this case, it might not be as important to have a precise temperature; however, as written, the concept of "cool" lacks precision. Instead of specifying a number (40°F, which in some cases recipes will have), they might either provide you a specific amount of time for the cooling to occur (a different way of quantifying the ambiguous value of "cool") or provide a standard of comparison such as "cool to the touch." The latter example is what we refer to as specificity, where the ambiguous term "cool" is provided a standard for comparison (other things you have touched and determined to be cool). This concept of specificity can be used for many ambiguous terms such as "young" or "quick," which are open to interpretation based on their use case. By providing a standard of comparison, you are improving the specificity of your statement without requiring a specific quantity to be provided in your instructions.

Imprecise: **Quickly** wash the samples **several** times in PBS.

Precise: Quickly wash the samples by **swirling the container holding the samples while filling it with PBS, then immediately removing the liquid** and repeating a total of **three** times.

Other ambiguous terms such as "better" or "best" may include an appropriate comparison modifier (i.e., the redesign was determined to be better than the provided reference design) yet still lack the type of precision needed for technical reports. Here, while a frame of reference is provided (the reference design), the sentence, as written, does not provide the writer's judgment criteria, that is, what they were looking for to determine what is "best." Assuming the criteria are not presented elsewhere in the report, it is imperative to include that information such that the reader can reach the same conclusion as the author.

Imprecise: The redesign was determined to be **better** than the provided reference design.

Precise: The redesign used **30% less energy and was 25% lighter** than the provided reference design.

The last example of precision that we want to emphasize is in the area of accuracy. Here we are talking about making statements that accurately represent your level of confidence or certainty on the subject at hand. Unless you are explicitly offering a guarantee, writing your findings or conclusions in terms of absolutes is often discouraged, as in many cases, your testing is being done on a finite number of specimens. This is not imprecision related to ambiguity; instead, it is making an adjustment to your language that recognizes the fact that any experiment, regardless of how well it is designed, includes a certain level of ambiguity.

Imprecise: The testing **proves** the design will be able to withstand the loads required for the structure.

Precise: The testing **indicates** the design will be able to withstand the loads required for the structure.

> **Engineering in Action**
>
> Break off into groups of three or four. Have one group member share a report they have submitted or (even better) plan to submit. Start by individually reading through the report and identifying opportunities to improve the sentence structure, conciseness, and precision. Then have each member share at least one suggestion with the group and discuss the pros and cons of the proposed change.

Storytelling

One way to think about report writing that may help you focus your efforts is to think about writing it much in the same way as you would write a story. Now we are not talking about Steven King's *The Stand* type of story, instead think about a simple children's story or perhaps a ghost story you would tell around the campfire. **Like those types of simple stories, your writing should have a clear beginning, middle and end with a focused plot that ties the entire thing together.** As you are writing your report, recognize where in the report you are (beginning, middle, end) and how the section you are working on fits into the overall plot and theme of the report. In general, this method requires that you plan out your report writing ahead of time, making sure you know what information needs to be disseminated to your audience. Some of this may be driven by the data you have collected, and as a result the figures you have generated. By developing an outline of the various sections that will be necessary for your report and having some idea of what information will be necessary within each section to drive the plot forward to its logical conclusion, you should be able to determine what information fits into this format, and what information is extraneous and can be removed.

Consistency

In addition to helping you focus your writing, outlines can be a good way to distribute effort when writing as part of a team. Many reports you will complete throughout your engineering career will likely be done as part of a team. Depending on how the group decides to divvy up the responsibilities, you might only write a small section of the overall report. For this type of writing exercise, it is extremely important to understand how your section fits into the overall report such that the final product is still a complete story, with a well-thought-out plot and an obvious beginning, middle, and end. In our experience, this is something that is easy to comprehend, but difficult to master.

When we talk about consistency throughout your report, there are multiple layers we are talking about. At the most basic level, there are superficial consistencies that need to be maintained throughout the report, such as cross-referencing material presented in previous sections as well as figure, table, and equation numbers. Appropriately referencing materials within the text of your report and the use of references and citations (both in terms of the style and formatting of the reference list and in-line citation as well as the actual references used) are also important to keep consistent. At a deeper level sits the understanding of where each section resides in the overall order of the report and what the consequences are of that order. Here, consistencies include ensuring the proper voice and tense are maintained throughout the document, that necessary information is presented in an orderly fashion such that the report can be easily understood by the reader, and that the information is not unnecessarily duplicated across sections written by different individuals. Additionally, any results or conclusions drawn at the end of the report need to be fully supported by information provided in earlier sections of the report. Some of the most obvious violations of this expectation are when a results section presents one set of information, but the conclusion directly contradicts these findings. When read separately (in the case where an individual is proofing their own work), the section is solid, but when considered holistically (as is the case when the instructor reads and grades the report), these contradictions become glaringly obvious. To avoid missing this deeper level of consistency, in addition to creating and strictly adhering to a report outline, all parties participating in the writing of the report need to be actively engaged throughout the writing process and need to take responsibility for the overall report quality, regardless of how sections were assigned to individual members. This higher-level consistency check may also be accomplished by a single individual taking on the role of an editor or proofreader. However, the final report still needs to be checked and verified by all group members to ensure errors have not been missed and that the intent of each section has not been compromised through the editing process.

TYPES OF WRITTEN COMMUNICATION

As you progress throughout your academic and professional careers, you will engage in a variety of different types of written communication. Listed below are examples of different types of written communication engineers commonly engage in:

- Letters, email, and memos
- Design reviews
- Proposals
- Test reports
- Project notebooks
- Operating manuals
- Design specifications
- Codes and regulations
- Documentation of computer code

Each of these types of documents will have specific requirements, and even within each document type, variations may appear from company to company. As such, we feel to provide full examples of each of these types of documents is beyond the scope of this text. Additionally, much of what we cover here in this chapter can be directly applied to these different genres. However, we do want to briefly touch on two types of documentation that we see students struggle with and provide you with some things to consider (including formatting) as you continue with your academic and professional career. Additionally, we want to discuss document content and expectations and will do so using standard lab reports as an example.

Letters, Email, and Memos

This genre is likely something you are already familiar with and, outside of perhaps test reports and project notebooks, have the most experience in creating. First and foremost, let's look at the three types of communication we have grouped together here, and describe when each should be used.

Letters and memos were certainly the de facto standard in business communication practices prior to the invention and adoption of email. The main difference between the two focuses primarily on your audience. Letters are most often reserved for correspondence that occurs outside your organization, although they can be used inside the organization for formal or non-routine activities. Memoranda (shortened here to the more commonly used term memo), on the other hand, are almost exclusively used "in house" to communicate a variety of types of information within an organization. Memos are commonly used to communicate information to employees in a direct and concise fashion. They typically ignore general introductions that you might include in a letter, as it is expected that the reader has a level of familiarity with the topic at hand that makes such information redundant. They are typically formatted into three very specific sections, which we will detail further below.

The most recent addition to formal business communication is email. Originally invented in 1971, electronic mail (shortened to email in the late 1970s) became prominent in personal and business communications in the late 1990s.[3] Today, email is commonly used both within a company as well as for communication to outside constituents; as such, its formality is often dictated by both use and audience. Although newer forms of instant communication platforms such as Slack and Microsoft Teams have emerged over the last few years, and whose popularity surged in 2020 due to the worldwide COVID-19 pandemic, email is likely to be one of the primary methods of interpersonal communication you will use throughout your professional career.

When it comes to each of these forms of communication, there are several things to consider as you draft your correspondence. As we have described in previous sections, understanding the level of professionalism required of each type of communication is perhaps the most important thing you can do to improve your writing. As faculty members, we routinely see email from students that is too informal in both their tone and the language used. While the format described below may vary depending on the corporate culture in your particular organization, it is better to err on the side of being more formal than expected and subsequently taking cues from your supervisors and co-workers to reduce your formality level. Each of the types of communication should include the following components: salutation, introduction, body, conclusion, and signature.

Header/Salutation

Of the three documents, the letter is the most formal document you craft. As such, there are specific sections that should appear in your letter. Regarding the header and salutation, as shown in Figure 5.1, these would include the date, followed by information regarding the recipient (name, address, etc.). Following that, you would include your salutation, which should be formal in nature and appropriately address the recipient (Dr. Adams, Mr. Jobs, etc. unless you are a close friend of the recipient), followed by a colon. For most business correspondence, letters would be printed on the official company letterhead, which often will include a return address.

November 12, 2018

Dr. Jessica Adams, CEO
Engineering Solutions, Inc.
4254 Innovation Way
Austin, TX 78709

Dear Mr. Adams:

FIGURE 5.1 Information to include in the header of a business letter

---

[3] https://www.theguardian.com/technology/2016/mar/07/email-ray-tomlinson-history.

In comparison to a letter, since a memo is often sent for internal use, the specifics regarding the address of the recipient are not necessary. Similarly, a formal salutation is also not necessary, as that information is contained within the header information, as shown in Figure 5.2. Note that in the header, the word *Memorandum* has been bolded, and is larger than the other text. Additionally, the header text has been aligned for readability, and there is a line separating the header from the message. The subject line is short but specific to what information will be included in the message.

Memorandum

TO: Michael Jobs
FROM: Dr. Jessica Adams
DATE: November 21, 2019
SUBJECT: Revision to Policy 3.032

FIGURE 5.2 Information to include in the header of a memorandum

You may have noticed the similarities between email and memos when it comes to the header information. The difference between the two is that much of the header information in an email is automatically generated by your email program, so you are not consciously adding it to your document. The one line you are adding (other than to whom the email is being addressed), which is the first thing someone will notice about your email, is the subject. As you would for a memo, your subject should be short and specific to the topic being covered in the email. Avoid ambiguity ("Hey there" or "Question" are generally not good subject titles), and **never** leave the subject line blank. Email also draws from letters in that you should include a salutation. Here, the salutation can vary depending on the situation as well as the culture of your company, but it is always best to err on the side of being too formal versus being too casual. In Figure 5.3, we use the same header information as we did in Figure 5.2 for the memo and show two different potential salutations based on how well we know the person we are emailing and if the person is inside or outside our company.

Introduction

The introduction should lay out the overall intent of the document. In a cover letter accompanying a report, or in a memo or email, this could be a short as a single sentence. "Enclosed are the requested reports from our last quality assurance audit on 12/10/20." For other types of letters, such as a cover letter for your resume or an introductory letter, this section might be two or three sentences. In short, this portion of your correspondence should answer the question, "Why should I read this?"

a.	TO: Michael Jobs FROM: Dr. Jessica Adams DATE: November 21, 2019 SUBJECT: Revision to Policy 3.032 Dear Mr. Jobs:	b.	TO: Michael Jobs FROM: Dr. Jessica Adams DATE: November 21, 2019 SUBJECT: Revision to Policy 3.032 Hi Michael:

FIGURE 5.3 Information to include in the header of a business email. a) is an appropriate salutation to a person outside your company that you do not know well, while b) might be an appropriate salutation for an email sent inside your company to a colleague you know well.

Body

The body should provide the details necessary for your correspondence. Again, depending on the type and scope of the document, this could be as short as one or two sentences or as long as a paragraph or more. A good way of organizing details in any type of writing is to start with general information, then drill down to more specific facts and details. Make sure you understand your audience and appropriately include information as needed, being as concise as possible and avoiding the inclusion of unnecessary information. In short, this portion of your correspondence should answer the question, "What do I need to know?"

Conclusion

Depending on the type and length of your correspondence, this section may contain up to two separate but equally important pieces of information; an overall summary or restatement of your main points and (possibly) a call to action. For shorter documents (i.e., less than one page), a summary statement is likely not necessary. For longer documents, this is a good way to circle back to your main point, making sure that in doing so you have remained focused on your purpose throughout the correspondence. This section may also include a call to action, or in other words, explaining what next steps are necessary to continue moving the described action forward. It is quite possible that due to the nature of your correspondence, neither of these pieces of information are necessary, and as such a conclusion might not be required. This section can often be used to answer the question "What do you want me to do?"

Letters, memos, and emails typically answer three questions a reader might have: "Why should I read this?" "What do I need to know?" and "What do you want me to do?"

Signature

For both emails and letters, you should include some basic information in your signature, including some form of complimentary closing (Sincerely, Best Wishes, Regards, etc.). At a minimum, this section should include your name and title (or in an academic setting, possibly your student ID), although for emails, your signature profile will (should!) also include appropriate contact information (email, phone, address, etc.). In general, the complimentary closing will be more formal for a letter than it will be for an email, but again, the type of correspondence and corporate culture will help dictate what is expected of you as an employee. Memos typically do not include a signature, although the sender of the memo may initial the document by their name prior to distribution.

> **Engineering in Action**
>
> Open your email sent folder and look over the last ten "professional" emails you sent. How many have appropriate subject lines? Can you revise any of those subjects to improve clarity? How many follow the salutation, introduction, body, conclusion, and signature format? Make a concerted effort for all of your professional emails over the next week to meet the layout discussed here.

Project (Lab) Notebooks

As with email, notebooks are likely something you have had experience with, perhaps in a science course or other engineering class. As with other forms of documentation described in this chapter, you will find variations in expectations amongst different supervisors, instructors, and companies. That being said, there is a fundamental purpose to keeping a record of your projects and ensuring that the record is preserved in a way that remains permanently accessible to your institution. As such, the guidelines we will discuss here may not be fully compatible with your company's method of recording your project efforts; however, it is likely that the methods they use achieve a similar outcome. In this section, we will focus on notebooks from the perspective of working in a laboratory setting. Recognize that in many ways, a project notebook will behave in a similar fashion, albeit sometimes with a slightly different focus.

The main purpose of documenting your work in a laboratory or project setting is to record the original intent of an experiment, project, or meeting, and in the process preserve all data, observations, and outcomes for future reference. If done correctly, this record should allow you to quickly find important outcomes associated with your entry, regardless of how far in the past the entry was made. Additionally, the record should allow others to understand your intent, interpret your results, and accurately replicate them if necessary. The need for meticulous records routinely comes up in furthering research and development efforts, demonstrating appropriate steps were followed during a regulatory audit, and validating data for quality assurance. As such, it is pertinent that a careful record is methodically crafted every time you create a new entry. While this may sound daunting, including the appropriate sections in your entries goes a long way in ensuring your notebook will live up to this responsibility. For a laboratory notebook, each entry should contain (at a minimum) the following sections:

- Date and title of the experiment
- The objective of the experiment (as originally intended)
- A detailed description of the experiment
- Any data or other outcomes and/or conclusions associated with the experiment

Within these sections, all pertinent facts associated with the experiment should be documented, including the equipment and materials used, any external conditions that might be important (for example, the temperature or relative humidity in the room might be an important condition to record), and the actual timing of the experiment (while the protocol might expect you to leave a reagent on your sample for *four minutes*, it might be important that when you ran the experiment, it actually remained on your sample for *five minutes*). These entries should be completed as you are conducting the experiment, or as close to when the experiment was conducted as possible to ensure your account of the activity is not biased due to recall. Overall, your entry should contain enough detail such that someone other than yourself could read the entry, recreate your experiment and end up with the same results (plus or minus experimental error).

As you are recording your findings in your notebook, you may be able to tell early on whether or not your results support your initial hypothesis or expected outcome. Many students become discouraged when faced with negative results and subsequently do not as enthusiastically record their findings as they might during an experiment that appears to be supporting their initial hypothesis. For an accurate record to be established, both positive and negative results should be thoroughly recorded. As has been shown throughout history, failed experiments can provide insight and leads for new experiments and conclusions. Many failed experiments have resulted in products that are wildly popular today, including Post-it Notes and Corn Flakes.[4] Additionally, recording failed experiments in detail can help ensure that future testing does not repeat these same mistakes. Nothing is worse than getting halfway through what you think is a new process, only to realize it was the same one that did not work previously.

Document Content and Meeting Audience Expectations

Earlier in this chapter, we described how it is important to treat your documents the same way you would treat a story, making sure your report has a well-defined beginning, middle, and end.

---

[4] https://science.howstuffworks.com/innovation/scientific-experiments/9-things-invented-or-discovered-by-accident.htm.

The text included within the report should support your story's plot and avoid superfluous content that does not significantly contribute to moving the plot forward. As you look through the various types of documents we described earlier in this section, you should be able to see how each of these documents might have a different story plot, and as such, contain slightly different content. Based on the document you are creating and the expectations of your audience (be it a faculty member, supervisor, etc.), you must ensure that your document contains all the elements required to meet those document expectations.

As an example of what we are describing, let's take a look at something you are probably familiar with; the requirements of a lab report. Most reports will contain at least four distinct sections: an introduction/background section, a materials and methods section, a results section, and a discussion/conclusion section. For some reports, you might be required to include additional information; for others, some of these sections might be omitted or separated (for example, you might be asked to include both an introduction and a background section). **Understanding which "sections" are required for your document will help you plan your document outline. Understanding what will be needed within each section will help you develop your story and determine how each section will contribute to your story plot.** In some cases, what is required in your report might be explicitly expressed by your instructor or supervisor either based upon assignment requirements (as you might receive as part of a grading rubric or assignment handout), direct conversation (descriptions given by your supervisor), or historical norms (observed when reviewing previous reports). Making sure these expectations are an explicit part of your document outline and that all sections are appropriately completed as part of your final documentation is a minimum obligation in fulfilling the requirements of your audience.

As you likely have experienced, simply including all the required sections of a report does not inherently ensure that the expectations of the audience have been met (but it is a good start). Many reports submitted in an academic setting will be assessed against a rubric or some other grading instrument, where different aspects of your writing mechanics, information literacy, and critical thinking skills will be assessed. You might have access to these rubrics in your classes, and if you do, it should become an integral part of your document outline to ensure those expectations are appropriately met. After graduation, although you will not receive a "grade" on your report submissions, your supervisors will still have expectations regarding writing mechanics, information literacy, and the clear expression of your critical thinking skills within your written documentation. Your ability to meet their expectations will likely play a crucial role in maintaining your current position within your company as well as your potential future progression. Adhering to the guidelines we described earlier in this chapter, as well as Chapter 4 on information literacy, is a good starting point. Additionally, adhering to any guidance provided by your instructor or supervisor is another excellent resource to follow. General written communication rubrics, such as the Written Communication VALUE Rubric from the Association of American Colleges and Universities (AAC&U),[5] can provide

---

5 Association of American Colleges and Universities, "Written Communication VALUE Rubric," 2009.

you with some additional guidance as to what undergraduate faculty expect in well-constructed written communication, which often will appropriately translate to post-graduate activities. Including some of the expectations contained within these types of rubrics might provide you with some additional guidance should you find or feel that your guidance is lacking. Additional guidance could also be gained through your university's writing center (or the equivalent). As discussed in our Engineering Success chapter (Chapter 2), taking advantage of opportunities and support provided through different university centers and services is an excellent use of existing resources.

DEFINING VARIABLES AND ABBREVIATIONS

In many forms of technical writing, variables and abbreviations (including both initialisms and acronyms) are often used to shorten technical language and names, thus making your text easier to read. As we described previously in this chapter, you should strive to make your technical writing as accessible as possible, using language that can be understood by as wide of an audience as possible. This does not mean that the meaning of your report needs to be simplified; it simply means that, when possible, reducing the complexity of your prose will lead to improved comprehension and retention. One way to reduce the complexity of your prose is the appropriate use of symbols, abbreviations, and acronyms to replace commonly used variable names and phrases. The key here is the word *appropriate*, as the use of any of these can easily make a report more difficult to read if they are ill-defined or used inappropriately. For the purposes of this section, we will describe our suggestions for abbreviations. Please recognize that the same rules equally apply for both acronyms as well as symbols.

Reducing the complexity of your writing leads to improved comprehension and retention.

To avoid using an ill-defined abbreviation in your report, you need to define it within your text the first time you use it. For example, if we were fabricating an object out of *polyether ether ketone*, we would likely want to use the well-known abbreviation PEEK in lieu of typing out "polyether ether ketone" each time we reference the material in our report. The first time we mention "polyether ether ketone," we would include the abbreviation in parenthesis. After introducing the abbreviation, we do not need to continue to use its full name in our report.

Polyether ether ketone (PEEK) was selected for this project due to its mechanical properties and biocompatibility. To improve osseointegration, the PEEK material was impregnated with hydroxyapatite (HA).

One exception to this rule is when you introduce an abbreviation within an abstract or summary statement that is included as a part of your report. In this case, you would also want to reintroduce the abbreviation within the main section of your text the first time the abbreviation is used. Additionally, including your abbreviations in a summary table either as a preamble to the report or in an appendix does not relieve you from needing to also include the abbreviation the first time you use it in the main text of your report.

Beyond avoiding ill-defined abbreviations, you also want to make sure that by including an abbreviation, you are not actually making your document harder to read.[6] **Be judicious in your use of abbreviations and choose those that are commonly used and will be easily recognized by your readers.** To do this, make sure you are following established standards as much as possible; for example, do not use a symbol typically reserved for shear stress (τ) to represent normal stress in your report.

INCLUDING FIGURES, TABLES, AND EQUATIONS

In many of your reports, it will be important to include figures, tables, and equations to better explain your findings or demonstrate your results (in fact, we suggested earlier that figures might be one of the tools used to help craft your document outline). We are sure you have all heard the adage regarding the value of a picture relative to words (arguments abound as to how many words a picture is worth, as well as the origin of the adage[7]), but the point is a well-constructed figure, table or equation can quickly describe something that might otherwise take a large amount of text. In terms of things to consider when adding these items to your report, in this section we will focus on the written aspects of figures, specifically discussing figure and table legends, axis labels, and table formatting. We will also discuss how figures, tables, and equations should be properly integrated into the main text of your report and what type of information should appear immediately before and after where these items are inserted into your report. We will start with equations, followed by tables and figures, noting where our suggestions overlap across multiple genres.

Equations

For many technical reports you prepare, you may find yourself needing to reference mathematical equations used in various calculations. When including these equations in your documents, it is important to label your equation in such a way that you can reference it within your document, as well as adequately define all variables, constants, and unknowns within the equation. As we also suggest for figures and tables, equations should be referred to in the main text of your document prior to their appearance and should be labeled in the order in which they appear.

---

6 https://www.psychologicalscience.org/observer/alienating-the-audience-how-abbreviations-hamper-scientific-communication.

7 https://www.phrases.org.uk/meanings/a-picture-is-worth-a-thousand-words.html.

For example, if you were describing the calculation of the volumetric flow rate of a fluid using a Venturi meter[8] with known cross-sectional areas A_1 and A_2 as well as a known pressure drop ΔP, you might do so as follows:

Equation 1 shows how the Venturi effect allows for the calculation of volumetric flow rate (Q) by using the observed pressure drop across the meter (ΔP) and knowing the cross-sectional areas of the larger (A1) and smaller (A2) sections.

$$Q = \sqrt{\frac{2\Delta P}{\rho}} \frac{A_1}{\sqrt{\left(\frac{A_1}{A_2}\right)^2 - 1}} \qquad \textbf{Equation 1}$$

In this equation, ρ is the density of the fluid being passed through the Venturi meter, which was measured to be 0.997 g/mL.

There are several things we want to highlight as items we feel are important in including an equation in a technical document (much of what we will cover here is equally valid for tables and figures as well). First and foremost, the equation is called out in the document prior to its introduction. This is important as it tells the reader to look for the equation. Without this reference, the reader may not know how or why the equation is important within the context of the overall report. Additionally, we reference the equation by name (Equation 1) rather than relying on its position (as shown below, or similar text). This could be especially important in situations where your document's layout and typesetting will be handled by someone other than you (for example, if you are sending a document to a publisher), and in the interest of page layout, the equation gets moved. **However, it is a good practice to refer to the equation by its name regardless of whether or not additional editing will occur on your document.** Also, note that any time we reference this equation, we capitalize the word "equation" and follow it by the appropriate number. This is true for tables and figures as well (as discussed later) and demonstrated throughout the book. When we speak of equations (or tables/figures) in general, the capitalization does not occur, only when referencing a "named" item.

Another important feature of this example is how all the variables, constants and unknowns were defined either immediately before or immediately after the appearance of the equation. Without these definitions, the reader may not fully understand how the equation works or what pieces of information are required for the calculations. For subsequent equations using the same variables, it is good practice to remind the reader what those variables are, but it may not be necessary to describe the variable in as much detail as you do the first time it is used.

---

8 https://www.omega.com/en-us/resources/venturi-meter.

There may be times where a mathematical expression does not warrant being included separately as a named equation and instead is better suited to be included in your document as an in-line equation or expression. The same rules as described above apply to this use case as well, as shown in the following example.

As seen in Figure 3, the suspension dampening from the car's shock absorber can be described by the equation $A(t) = A_0 e^{-kt} \cos(\omega t)$ where A_0, k and ω change the amplitude, frequency, and amount of dampening. For our spring system, the best fit curve resulted when $A_0 = 2.5$, $k = 1.5$ and $\omega = 3\pi$ when t was measured in seconds and the amplitude was measured in millimeters.

In this equation, we describe the equation in line with the rest of the text. As such, this equation has no label (i.e., number), however, all the unknowns or constants are described (and in this case defined). In writing the equation this way, it will not be as easy to refer to the equation later in the report, as such this method of including an equation is appropriate when you only need to refer to the equation in question a limited number of times (most often only once). Longer or more complex equations, particularly ones with numerous sub/superscripts or fractions, can be difficult to read in-line, and thus may need to be presented separately from the rest of the text, even if the equation in question is only referred to once in the report.

One minor point to consider when it comes to the inclusion of equations is the appropriate formatting of the equation. For very simple equations, using a word processor with the appropriate use of subscripts and superscripts may be adequate. For more complex equations, you might need to use an equation editor, which is often provided within a word processing program and can be used for both in-line as well as stand-alone equations. For very complex equations, you might need to consider using a typesetting program such as LaTeX[9] to ensure your equation formatting is appropriate, both in terms of layout as well as font size.

Tables

Tables, much like equations, should be appropriately numbered, directly referred to in the main text of your document, and appear within the document after the in-line reference but as close to that reference as possible. Tables (and figures) differ from equations in that their label should include both a name and a title. Additional information needed to understand the table after the title can be added but is not necessary. We will refer to this number and title as the table legend (not to be confused with what most plotting software will consider a "legend", which we will refer to as a chart legend, and will discuss in greater detail in the section on Figures), an example of which follows.

---

9 https://www.latex-project.org/.

As shown in Table 4, the number of cycles to failure varied as a function of the load applied to the material as well as its annealing temperature.

TABLE 4 Effect of Annealing Temperature and Applied Compressive Stress on Fatigue Failure in Alumina-SiC Nanocomposites

Annealing Temperature (°C)	Normal Stress (MPa)	Cycles to Failure
1200	100	1.25×10^6
	150	9.50×10^5
	200	8.70×10^5
1400	100	2.11×10^6
	150	1.85×10^6
	200	1.05×10^6
1600	100	2.65×10^6
	150	2.05×10^6
	200	1.80×10^6

There are two important things to note in regard to the table as shown. First and foremost, note that the table legend appears at the top of the table. This is different from figure legends (discussed later), which will appear below the figure. Note that this is not a universal expectation, so it is best to talk with your instructor or supervisor to determine their specific format expectation. As we described in the previous paragraph, the legend includes the table number as well as a descriptive title. Additional information such as statistical information, symbol identification, or other information that would aid the reader in understanding the table would also appear in this legend as sentence(s) separate from the title. In essence, the table legend allows the table to "stand alone"; that is, the information contained within the table can be understood without needing to read the rest of the report.

The second item we want to focus on in regard to the table is the table format. Here, we are not talking about the aesthetics of the table but instead want to focus on the content of the table column titles and the formatting of the data within the table. Note that the column titles have units associated with them and that these units *only* appear in the title row. Additionally, note that the data in each row has a similar number of digits, and for large numbers (such as our cycles to failure column), how scientific notation was used. We will discuss the use of significant figures in a later section in this chapter, and significant figures play a large role in how many digits appear in each column, but keeping those numbers consistent within each column and making sure they are easy to read (scientific notation) is very important when creating an appealing, easy to read table. One other way to address large numbers is to include scientific notation in the title, which would indicate that all the numbers listed in that column are represented by the number in the table multiplied by the scientific notation contained in the title. A variation of our example table with this type of representation is shown in Figure 5.4.

Annealing Temperature (°C)	Normal Stress (MPa)	Cycles to Failure (x 10^6)
1200	100	1.25
	150	0.95

FIGURE 5.4 An example of a table using scientific notation in the column title

Figures

Figures share many of the same legend characteristics as tables in as much as their legends should minimally contain an appropriate figure number as well as a descriptive title. Additional information such as statistical information (p-values, number of replicates per condition, etc.) or chart symbols/notation (describing what the bar lengths, error bars, and/or symbol shape represents, etc.), if not directly included in the figure, should appear here. As with a table, the figure, along with its legend, should "stand alone": that is, the reader should be able to interpret the figure information without needing to read the rest of the report. **One difference from table legends is that figure legends appear below the figure in question.** As with tables, the figure should only appear in the report after being referenced in the text and should appear as close to the in-text reference as possible. While it is most common to reference the figure as "Figure #" (the same as we did with tables), it may be acceptable to shorten the reference to "Fig. #" (there is no similar abbreviation for tables).

Within this section, we will cover two types of figures we see quite often in reports—bar graphs and scatter plots. Bar graphs are useful in situations where the data can be organized into groups and you are interested in comparing a singular output across those groups. Examples of this would include measuring the mechanical tensile strength of different materials, the maximum heart rate for different groups of individuals when exercising, or the volume of product produced using different types of chemical reactions. Scatter plots are useful when tracking data over time or when the data is dependent on another variable. Examples of this would include measuring the maximum temperature in your garden each day (temporal variation) or how moisture content in the soil varies after a specified period of time due to differences in humidity (moisture content is the dependent variable and humidity is the independent variable).

Bar Graphs

Bar graphs are commonly used to compare a singular data type or measurement across multiple groups or cohorts. The magnitude of these graphs is often used to indicate the average value of the measurement within the cohort, and the error bars either the standard deviation or standard error within the cohort. Programs such as Excel can calculate these values directly (in Excel you can use the "=AVERAGE(A1:A25)" and "=STDEV(A1:A25)" commands where "A1:A25" indicates the range of cells that contain the data you are interested in plotting), so there isn't necessarily a need to calculate these values manually (although you certainly could if you wanted to). You can then plot these values in Excel using the the chart function and selecting a bar graph.

When you do this, you will likely find that the default chart Excel generates does not meet the expectations of your report standards. Even based on the information we've provided in this section, the default figures generated by many spreadsheet software packages (such as Excel) will not be sufficient. One thing we encourage students to immediately look at and consider modifying are the axis values and labels. If you are applying our figure legend recommendations, your figure title should appear in that legend; as such, you do not need to include a figure title in your actual figure. This is something that Excel, for example will include by default. As shown in Figure 5.5 this should *not* be included.

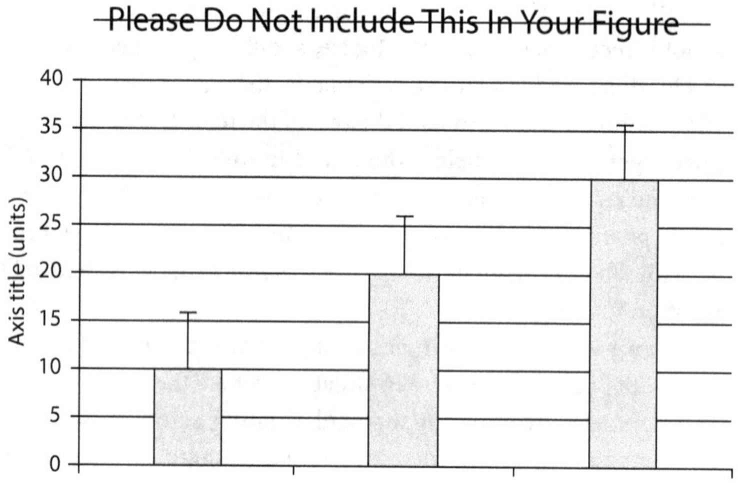

FIGURE 5.5 Example of required elements for an appropriate figure.

In addition to not including the title in the figure itself, as the title is included in the figure legend, you may need to add additional formatting elements to the default graph created by your software package. This might include adding axis titles, adding or subtracting chart legends (we will discuss those more when we talk about scatter plots), changing the default font size of your axis titles and/or your axis units, changing the default color of your bar graphs, changing the thickness and/or type of your error bars (standard deviation or standard error), et cetera. Note the information included in Figure 5.5 includes appropriately formatted values for the x- and y-axis as well as x- and y-axis labels, which should include the units associated with the axes values as necessary (in our example, both the y-axis and x-axis title includes units, in some cases you might need to include units for one axis but not the other). When we describe the y-axis values as being "appropriately formatted", what we mean by that is that we have chosen appropriate units for the y-axis such that we have numbers with less than 3–4 digits along the y-axis. If our example was the yield stress of various materials, and Example 1 had a yield stress of 10,000 kPa, we would choose to make the unit scale of the y-axis MPa so that our graph reads 10 MPa (the equivalent of 10,000 kPa). To increase the readability of the figure, you might need to increase the default font size of both the axis values and the axis labels (with labels typically maintaining a larger font size than the font size of the values).

When making a figure panel, where more than one chart appears in the same figure, you want to make sure that the axes values are the same across all the figures in the panel so direct comparisons can easily be made. As shown in Figure 5.6, allowing the y-axis value to vary

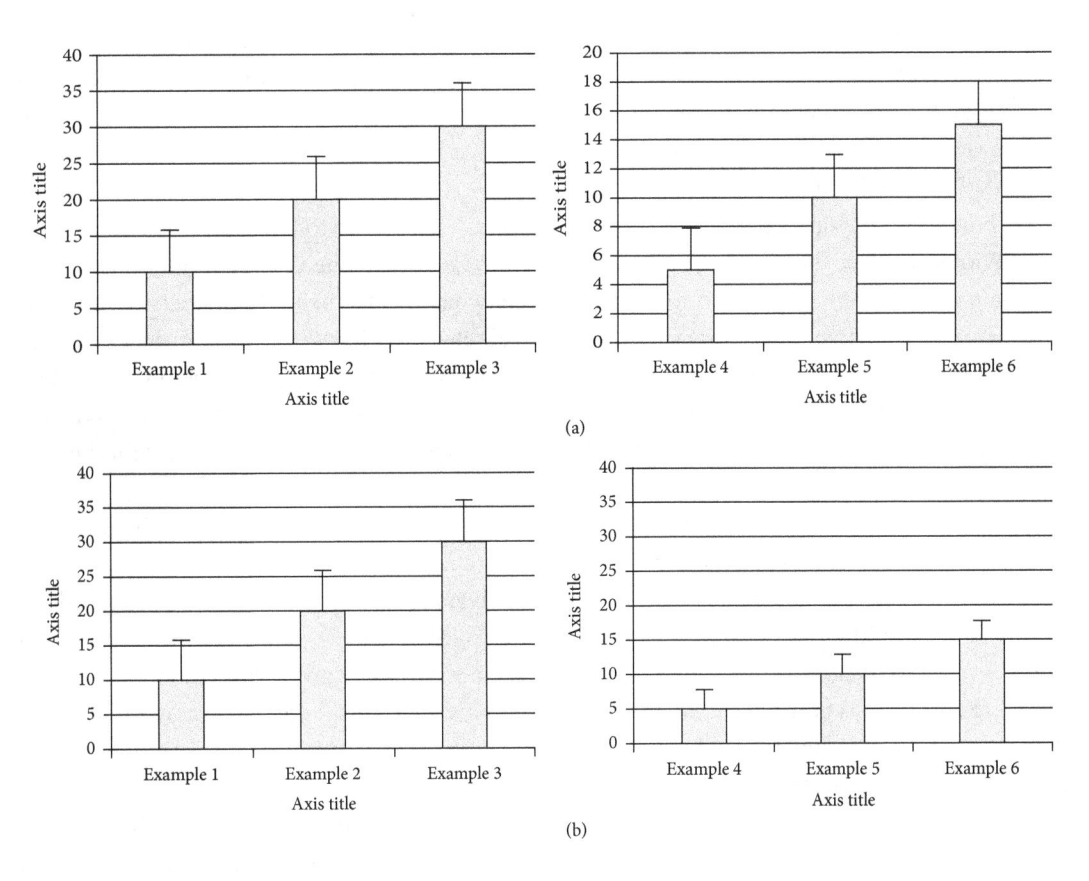

FIGURE 5.6 Examples of panel figures including multiple charts. a) The left and right panel have different y-axis values, making comparisons between the two example groups difficult. b) The left and right panel have the same y-axis value, making comparisons between the two example groups easier to visualize.

between the two charts (panel a) makes Examples 4, 5 and 6 appear to have the same magnitude as Examples 1, 2 and 3 respectively, even though in reality (as can easily be seen in panel b) Example 5 has the same magnitude as Example 1.

In addition, the font size has been increased in this panel for both the axis values and titles, as well as the error bar thickness to improve readability (in panel a, the font and line thickness was kept the same as found in Figure 5.5.)

Scatter Plots

Many of the suggestions that are described in the section regarding bar graphs are relevant to scatter plots, specifically those regarding titles, fonts, colors, et cetera. One of the main differences between scatter plots and bar graphs is that you are plotting a dependent (y-axis) and independent (x-axis) variable, thus your data will be represented as either a singular point or perhaps an average point (with appropriate error bars) depending on the type of information you are attempting to represent. Let's take a look at two different types of scatter plots that examine the daily average temperature in Fort Myers, Florida over time (note that this data comes from the Florida Climate Center at Florida State University[10]). We'll start with the scatter plot shown in Figure 5.7, which demonstrates how the mean daily temperature in Fort Myers varied in the month of January from 2018–2021.

Although this data may not be particularly useful as a predictive tool nor does it demonstrate any obvious patterns, there are multiple features/efforts made that need to be highlighted. Many elements of this graph were customized to generate what you see here. Axis labels were added to the x- and y-axis, font sizes were increased for all elements in the chart, and the shape and size of each set of data points was changed. We've included a chart legend below the graph, that indicates which symbols correspond to each year. Additionally, the span of each axis was customized (40 – 85°F along the y-axis and 1–31 days along the x-axis) to best represent the data in the graph.

One additional feature to note is that no line was added connecting the individual data points. Each point here represents a discrete measurement (actually the mean of all the data taken during that day) made at the airport in Fort Myers (Paige Field). Any type of line added to this figure would indicate there is a known relationship between those points, which there is not. If you were developing a model to predict the temperature on a given day, you might plot the model results on this graph as a line to demonstrate how well the model fit the actual measured data, but you would want to avoid connecting these data points, even if such a line would make differentiating between the various years more obvious to the reader.

The second scatter plot we want to present will still look at the mean temperature, but we have now calculated it as an overall mean and a standard deviation for each month and will look at that data for the years 2020 and 2021. This information is presented in Figure 5.8.

---

[10] https://climatecenter.fsu.edu/climate-data-access-tools/downloadable-data

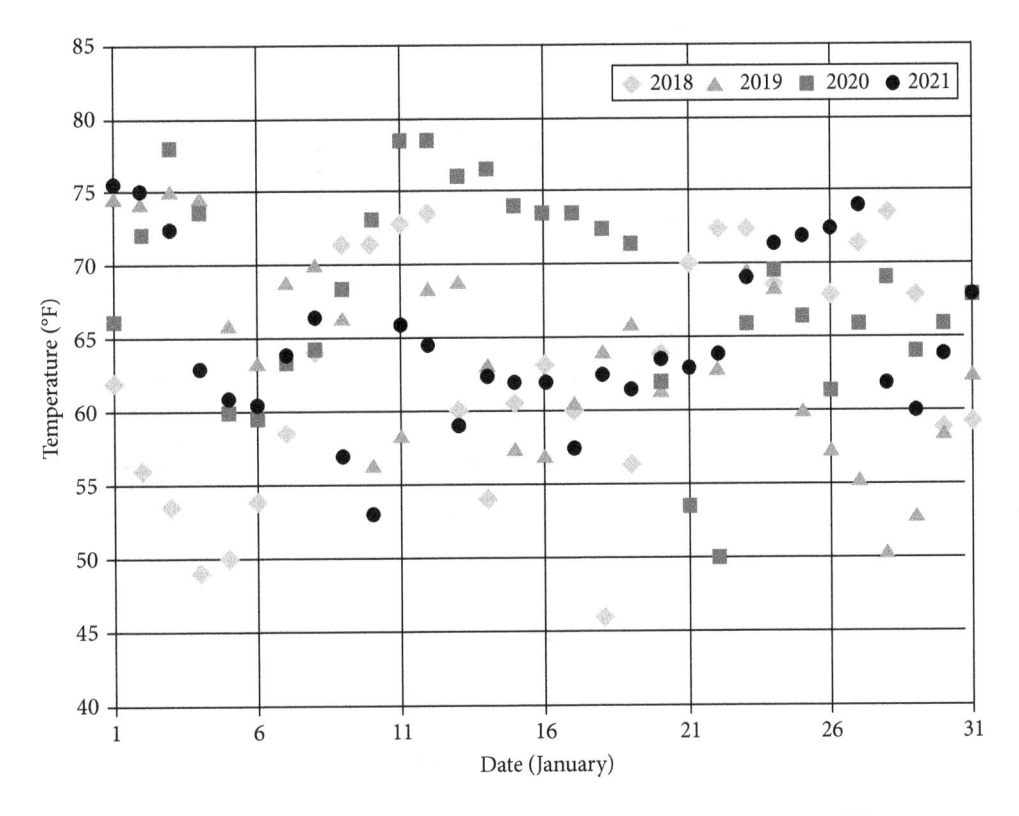

FIGURE 5.7 Daily mean January temperatures in Fort Myers, Florida from 2018–2021.

As before, multiple manipulations were needed to generate this graph, including additional titles for the axes, changing the font sizes of the text, changing the size, shape and color of the symbols, including the appropriate lengths for the standard deviations (and only using the vertical standard deviation as opposed to the horizontal standard deviation which might also exist in a scatter plot) and increasing the thickness of the error bars. We changed the axis label to represent months by the first letter of the month versus representing the month by a number (1–12). In this figure, we included the chart legend at the top right of the graph and boxed it in to better set it off from the rest of the graph. Like Figure 5.7 we did not include any type of line to connect these data points as that would either imply a relationship between the points or represent a continuous measurement of data. Note that in this figure we included both positive and negative error bars, whereas the error bars we used with the bar graphs (Figures 5.5 and 5.6) used only positive error bars. We recognize the data presented here could have just as easily been presented as a bar graph, but particularly with Figure 5.7 a graph with that many bars would have been extremely busy.

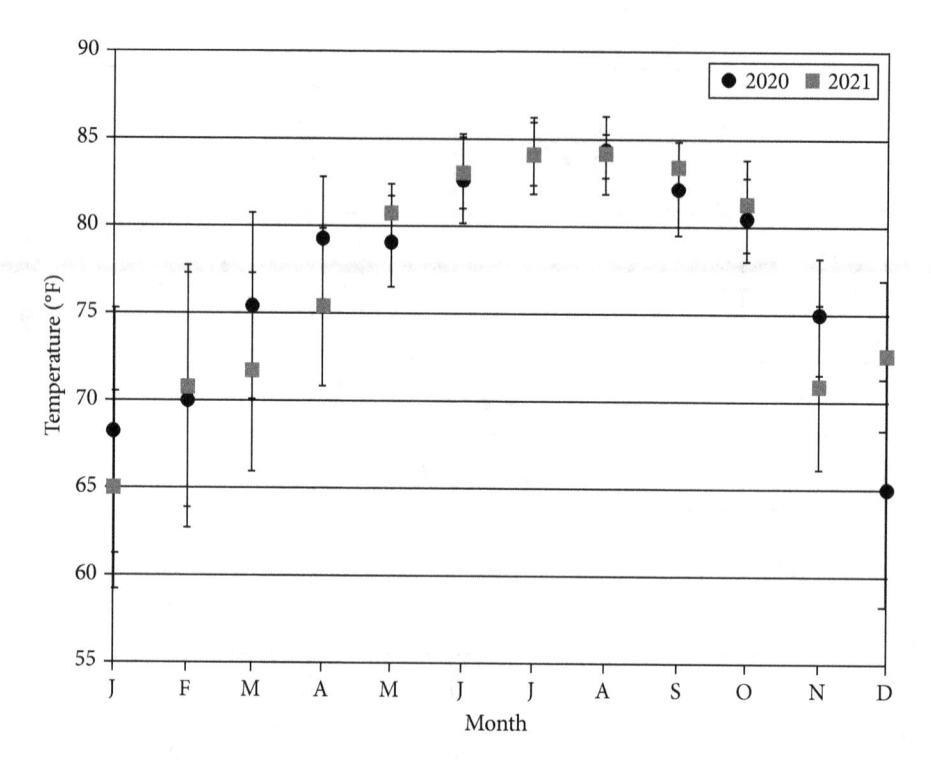

FIGURE 5.8 Monthly mean temperature with standard deviation for Fort Myers, Florida from 2020–2021.

SIGNIFICANT FIGURES

In several previous sections, we have described the need to use "readable" numbers, which in many ways goes hand in hand with (although it does not dictate) significant figures. We are assuming at this point in your academic career, you have discussed the rules associated with significant figures, and what they represent. As a summary, remember there are five rules to counting significant figures.

1. All non-zero figures are significant *(946 contains 3 significant figures)*.
2. Any zero between two non-zero figures is significant *(405 contains 3 significant figures)*.
3. Trailing decimal zeros are significant *(0.0**400** contains three significant figures which have been bolded)*.

4. Zeros prior to the decimal are significant if the decimal is present *(100. contains three significant figures while 100 contains one significant figure; 100. would commonly be written as 1.00 x 10² to avoid any confusion).*
5. Leading zeros **are not** significant *(in Example 3 (0.0400), the second zero (bolded) is a leading zero, while the first zero is added to improve the readability of the number)*

While these rules may seem somewhat arbitrary, they are much easier to understand if you have had the opportunity to discuss or use significant figures in the context of measurements. The number of significant figures indicates the precision of the measurement you are able to make. For example, if you were to weigh yourself on a standard electronic bathroom scale, that scale might read 170.5 pounds. In this instance, your scale's precision is limited to tenths of a pound. As written, this number has four significant figures. Alternatively, if you look at a standard ruler, on the metric side of the ruler, there are markings for tenths of a centimeter. As a general rule of thumb, you should be able to determine a value that is 1/10th the smallest division on the scale of your measuring device, and therefore you would be able to measure to 1/100th of a centimeter. However, because you are estimating the distance between those two markings, there is an uncertainty to your estimation that is equal to ½ the length of the smallest division, or 1/20th of a centimeter. Therefore, you might measure something to be 10.54 cm, which as our previous example, would have four significant figures, however, if you wanted to report both your measurement and your level of uncertainty you would express this length as 10.54 ± 0.05 cm. If you had a ruler that had markings 10 cm apart, your level of accuracy based on our general rule of thumb would be limited to 1 cm; thus, your answer might be 9 cm or 18 cm, which would have one or two significant figures, respectively. With your level of uncertainty, this would translate to an expressed length of either 9 ± 5 cm or 18 ± 5 cm.

As you may recall from an earlier science or math course, **mathematical operations involving numbers containing various levels of significant figures result in solutions that are only as accurate as the least precise measurement.** Therefore, your answer will only be as precise as your least precise measurement. With this in mind, it should be noted that exact numbers, such as the number of students in a classroom or defined numbers, such as the number of molecules in a mole or the number of carbon atoms in a molecule of glucose have an *infinite* number of significant figures. As such, these types of numbers do not impact the precision of the solution. In reporting your answers within your report, your values presented either in your text or in your tables and figures should appropriately adhere to the precision of the measurements made in your experiments.

> **Example:**
>
> **Question:**
>
> *Your measuring tape has markings for 1/10 of a centimeter. When you measure the area of your bedroom, you measure it out to be 289.80 cm by 345.52cm. How would you express the area of your bedroom in square meters using an appropriate number of significant figures?*
>
> **Solution:**
>
> First convert your lengths from cm to m:
>
> 289.80 cm = 2.8980 m (5 significant figures)
>
> 345.52 cm = 3.4552 m (5 significant figures)
>
> Area = 3.4552 m x 2.8980 m = 10.0131696 m^2 (output from calculator)
>
> Area = 10.013 m^2 (5 significant figures)

STATISTICS

One word we commonly see students use in their reports when describing substantial differences between two populations is the term "significant." As we discussed earlier in this chapter, it is best to avoid words that lack precision, such as large, short, or significant, and instead replace those terms with exact values or, at a minimum, a means of comparison. Instead of stating that group A was *significantly larger* than group **B**, state that group A was *twice the size* of group B. That being said, in many forms of technical writing, the term "significant" can and should appear in the text when it is referring to or describing groupings that have been demonstrated to be *statistically different*.

Although the intent of this book is not to provide you a full dive into statistical methods, we do want to take this opportunity to get you thinking about how statistics might impact your writing and recognize how it might impact your experimental setup and implementation. For those students that have already been exposed to statistics, either through a prior course or a priori knowledge, the next few paragraphs should be a review. For students who have not previously had exposure to statistics, hopefully these paragraphs will give you something to think about when you do take a statistics course as part of your undergraduate studies.

The application of statistical analysis to experimental protocols and using the outcomes of that data to verify current behavior or predict future behavior is something engineers do on a daily basis. From quality assurance to validation engineering to performance testing, all aspects of engineering rely on the appropriate use and interpretation of statistical methods. Common engineering and manufacturing concepts, such as signal to noise ratio, mean time to failure, empirical equations, predictive behavior, and sampling all are rooted in understanding how statistics can be applied to observed phenomena, and how that information can be used to validate or predict appropriate (or inappropriate) behaviors.

All aspects of engineering rely on the appropriate use and interpretation of statistics.

At its most basic level, statistics is defined as the science of collecting, analyzing, presenting, and interpreting data.[11] From an engineering perspective, this data could be anything from the output of a voltmeter, the load being applied to a test sample, or the absorption of a specific wavelength of light. As you can see from our definition, statistics plays an integral role in planning, executing, analyzing, and presenting any and all experimental data we collect. Using these tools, we can determine the variability of a system of measurements and whether an observed difference in separate populations is "real" or simply a result of the variation within each population. This is of particular importance if we are attempting to design an experiment to determine if object A performs better or worse than object B.

When you have collected data in a science or engineering course, you may have repeated the same experiment a handful of times to ensure you understand how the experiment is conducted and perhaps to provide you with enough data to verify that there are differences between object A and object B. As an example, suppose you are tasked with determining whether heat treatments such as hardening (heating the material to 910°C followed by rapid quenching in water) or tempering (heating the material to 450°C followed by cooling in room temperature air) improve the mechanical properties (e.g., yield strength) of steel.[12] In class, you treated three samples of each material (untreated, hardened, and tempered) and then tested them to failure on a material testing machine. In comparing the data from your samples, you might report the average and standard deviation of each material treatment and present that information as a bar graph where the bar represents the average value of your three samples and the error bars represent the standard deviation (similar to the format provided in Figure 5.5). These data are the **descriptive**

---

[11] https://www.britannica.com/science/statistics.
[12] M. F. Hasan, "Analysis of Mechanical Behavior and Microstructural Characteristics Change of ASTM A-36 Steel Applying Various Heat Treatment," *Journal of Material Science & Engineering*.

statistics of the samples tested, and the reporting of this information demonstrates both the central tendency (mean) as well as the variability (standard deviation) of your data sets.

Continuing with our example, one of the things we would want to write about in our report is whether or not either of the heat treatments improves the mechanical properties of our steel samples. In graphing the data, we may see that the average value of the hardened steel is greater than either the tempered or untreated samples. We could easily state that the average mechanical strength of our three samples of hardened steel is greater than that of the tempered or untreated samples but writing our results in that fashion only provides us with information regarding the specific samples we tested. For a manufacturer, this test would not give us enough information to determine if hardening the steel is worthwhile (assuming our use case requires the steel to be stronger than as cast [i.e., untreated]). To determine whether or not the additional time and cost associated with the heat treatment will reliably improve the mechanical strength of our steel, we not only need to look at the average value of our samples, but also understand how much variability there is from sample to sample. Our collection of data would need to be large enough to run an appropriate statistical analysis that would examine both the average value as well as the standard deviation of the two (or more) treatments being considered and determine, based on those values, the likelihood that those sample populations are actually different from one another (in our experiment with three replicates, there is not enough data to demonstrate that the differences shown in our graph are not simply due to random chance). In statistical terms, this idea is expressed as a **p-value,** and it represents the likelihood that observed differences are simply due to random chance. For many scientific publications, a p-value of less than 0.05 (meaning that there is a 95% probability that the observed differences are not due to random chance) is required to describe the difference as being statistically significant. This type of information is often included in reports in both tables and figures, where p-values less than 0.05, less than 0.01, and less than 0.001 will be shown as different symbols within the chart or on a graph. In many ways, these values can be thought of as confidence levels of good, better, or best. Understanding what types of tests need to be run to establish these values is beyond the scope of this book; however, by establishing such values in our experimental example, our manufacturer will have more confidence in our results being applicable to samples beyond our sample set and as a result, may end up changing their manufacturing processes such that all their steel ends up being heat treated.

Beyond establishing that observed differences are likely or unlikely to be due to random chance, statistics can also be used to establish trends and equations that predict future behavior. If you were to plot the force needed to elongate a spring to 1, 2, or 3 cm beyond its initial length, you would find that the results are roughly linear (theoretically, it should be perfectly linear, but there may be some variation due to experimental error). By determining the relationship between force and elongation, you can predict the force required to elongate the spring 8 cm beyond its initial length without physically needing to conduct the experiment. This ability to predict future behavior or develop a mathematical expression describing the relationship between two or more variables can be accomplished through **regression analysis**, something you may have used in Excel or another type of spreadsheet software when you plotted your data as a scatter plot and then asked Excel to add a trendline to your data. For our spring, a simple linear regression is sufficient, but higher-order polynomials or other relationships such as exponentials or power laws may be needed to describe more complex phenomena.

APPENDICES

At the start of this chapter, we suggested your story should remain clear and concise, and like any good story, it should have a clear beginning, middle, and end. However, as we introduced the concepts of equations, tables, and graphs, as well as the use of descriptive statistics and linear regression, you might begin to ask yourself how do you ensure that your story remains clear and concise with the overwhelming amount of data and information generated in a particular lab or study? When writing a report, it can be very easy to inundate your readers with a lot of additional information that may not be pertinent to the story you are attempting to tell. Additionally, your supervisor, client or instructor may require specific pieces of data or types of analysis that, while important, don't do much to move the "plot" of your story forward. For example, perhaps your instructor wants you to include tabulated data from each individual load-deformation experiment you completed when you tested the various heat treatments for your steel samples as described in the section on Statistics. This ultimately will require nine additional tables to be added to your report that likely do not significantly move your story forward. How can you best navigate through this seemingly divergent set of requirements and expectations?

One potential solution to these conflicting sets of expectations is the appropriate use of appendices. Many technical reports will include appendices for materials that are either important only to certain readers of said report, or perhaps help document certain aspects of the report that enable additional analysis of the data, but overall do not significantly change the report narrative nor does their omission from the main text of the report reduce the credibility of the report findings. As with many sections in this chapter, opinions may differ on how/when appendices should be used, but in many cases, appendices (similar to supplemental materials provided in some journal articles) can effectively bridge the apparent gap between specific report requirements (i.e., tabulated data from all experimental trials) and producing a clear and concise graph for your report (one single figure consisting of 3 bar graphs each representing the central tendency of one of the three heat treatments tested as well as error bars representing the variability of the trials). The graph is likely sufficient to tell the story you want to tell; however, some readers might be very interested in the tabulated data. As with figures, tables, and graphs, appendices need to be labeled (Appendix A, B, etc.) and referred to in the text. From our example, such a call-out might appear as follows.

As shown in Figure 1, the hardened steel demonstrated a 75% increase in tensile strength compared to either the tempered or untreated samples. Load-deformation data collected from each sample used in Figure 1 is included in Appendix A.

By definition, appendices are supplementary materials usually attached at the end of a piece of writing.[13] As such, unless explicitly told otherwise, they should be used to house additional materials that do not significantly impact the story your report tells. If you find yourself referring to a specific figure within an appendix multiple times, it is highly likely that the appendix is not the appropriate place for that specific figure. On the other hand, anything placed in the appendix needs to be appropriately referenced within the main text of your report, so a reader knows where to go should they be interested in those supplemental materials. Stylistically, an appendix is generally the last section(s) of a report (the *s* indicates you might have multiple appendices in a single document), coming after any bibliography or reference section.

CHAPTER SUMMARY

This chapter focuses on the key characteristics associated with effective written communication in a professional setting. Within this chapter, the following topics were covered:

- Suggestions and recommendations for improving *writing clarity*
- The importance of *outlines* and integrating *audience expectations* into written communication
- An overview of the different types of professional written communication as well as the various levels of professionalism applied to these documents
- How to appropriately include tables, figures, and equations, as well as variables, abbreviations, and significant figures in written communication
- A brief introduction to *statistics* and their use and importance in engineering
- A discussion on when appendices may be incorporated into a document and how these contribute to balancing the inclusion of additional material without disrupting the flow of your story

END OF CHAPTER ACTIVITIES

Individual Activities

1. Select a report you have due in the next week or so. If you do not have one due, you can apply these activities to a report that has already been submitted. Go through the report line by

---

[13] https://www.merriam-webster.com/dictionary/appendix.

line, reading aloud each sentence. Use the "track changes" option in Word, the "suggesting" option in Google, or a similar approach to edit the document for sentence structure, conciseness, and precision. Once you have completed your edits, review the original versus edited document side by side (or back to back) and answer the following questions:
   a. What are the primary differences between the two documents?
   b. Is there a particular type of edit that was most common?
   c. How does the flow of the revised document compare to the original?
   d. What are the top three things to consider when writing future documents?
2. Create a draft of a professional email for each of the following situations:
   a. You were doing well in a course and felt you understood the material, but midterms were recently returned, and you did not get the grade you were expecting. You want to send an email to your instructor asking what resources are available to assist you in improving your grade.
   b. You saw a flyer advertising a research opportunity on campus in the lab of a faculty member with whom you would like to work. The flyer asks for interested individuals to contact the faculty member via email. What do you write?
   c. In searching a job board, you find an internship position that sounds perfect for you. The listing asks for you to email the engineer to indicate your interest and to include your resume. Create this email.
3. Go to the website https://ourworldindata.org/ and pick a topic to consider. Within that topic, take a look at some of the figures that the site creates. Consider the following:
   a. How much information can be gained from the figures alone compared to the figures and text?
   b. What features are used to emphasize information within figures? What makes these approaches more or less effective?
   c. Why is a certain type of figure/graph used for a particular set of data? What might be another method or figure type that could also present that data clearly?
4. Download a data set from https://ourworldindata.org/ on a topic of your choice. Complete the following:
   a. Select a portion of the data and create a graph that demonstrates a comparison between at least three different data sets.
   b. Add appropriate axes labels, a chart legend, major and minor gridlines as needed, etc., to your graph.
   c. Copy and paste your graph into a document and add a label, title, and caption. Make sure that all the features of your graph are appropriately visible in the new document (i.e., font size, error bars, etc.).
5. Download a data set from https://ourworldindata.org/ on a topic of your choice. (This can be the same set as used previously or a new set). Complete the following:
   a. Select a portion of the data and create a table that demonstrates a comparison between at least three different data sets.

    b. Copy and paste your table from the spreadsheet into a document and add a label, title, and caption. Adjust font size and spacing to try and allow for the entire table to fit on a single page.
    c. If your table extends over more than one page, split the table appropriately.
    d. Create two versions of the table - one presenting a brief summary of the key data for inclusion in the main part of the document, and the second including the complete data set and more appropriate for an appendix.

Team Activities

1. Have each member of your group take a turn sharing a brief two-to-three-minute summary of their weekend. Once every member has shared, consider the following questions:
    a. What was similar about the WAY in which group members shared their stories?
    b. What were some of the differences in the WAY in which group members shared their stories?
    c. Whose weekend do you feel you know the most about? Why might that be the case?
    d. Did any story seem to go off track or have a TMI moment that may have been more appropriate as optional information?
    e. How does this activity relate to improving written communication?
2. As a group, decide on an activity that requires multiple steps to complete. Once you have agreed upon the activity, complete the following:
    a. As individuals (no communication between group members), create a twelve-step process (list of instructions) for completing the activity.
    b. Once all individuals have completed their list of instructions, divide each list by the number of individuals in the group (e.g., if your group has four members, your list will be divided 1–3, 4–6, 7–9, and 10–12).
    c. Rotate the divided instructions between group members (e.g., Group Member A should have their section 1–3, Group Member B's section 4–6, C's 7–9, and D's 10–12. All group members should have one section from each group member and have one of each section).
    d. Again, as individuals, take your new list of instructions (composed of sections from all group members) and edit it for consistency. Make sure to differentiate the edits from the original steps.
    e. Once all individuals have completed their edits, come together as a group and look at all the edited versions of the instructions. Consider the following:
        i. What might have caused certain group members to have to complete more edits than other group members?
        ii. If you knew ahead of time that the lists were to be divided, would this have changed how you approached your individual instructions? How or why not?
        iii. What could have made the editing process easier?

3. As a group, compare the emails you created as individuals in question 2 from the "individual activities" section. What similarities and differences do you notice between emails of the same scenario from different individuals? What similarities exist between emails from the different scenarios? Are there approaches that another team member took that the group finds particularly interesting or would use in future emails?

CHAPTER REFERENCES

Association of American Colleges and Universities. 2009. "Written communication VALUE rubric." https://www.aacu.org/value/rubrics/written-communication.

Banks, David. 2017. "The extent to which the passive voice is used in the scientific journal article, 1985–2015." *Functional Linguistics* 4, no. 1 (Dec.): 12.

Bly, Robert W. 1998. "Avoid these technical writing mistakes." *Chemical Engineering Progress* 94, no. 6: 107.

M. F. Hasan. 2016. "Analysis of Mechanical Behavior and Microstructural Characteristics Change of ASTM A-36 Steel Applying Various Heat Treatment." *Journal of Material Science & Engineering* 5, no. 2. Accessed January 4, 2021. http://www.omicsgroup.org/journals/analysis-of-mechanical-behavior-and-microstructural-characteristics-change-of-astm-a36-steel-applying-various-heat-treatment-2169-0022-1000227.php?aid=67903.

CHAPTER 6
ORAL COMMUNICATION

BEFORE YOU BEGIN

Take a minute to consider the following questions:

- What qualities or characteristics does a good public speaker possess?
- How do you feel about giving a public presentation?
- How do you typically approach preparing for an oral presentation?
- What similarities might a technical or professional presentation have when compared to someone performing stand-up comedy or improv?
- Consider a dynamic speaker you observed recently. What made the individual particularly interesting or engaging?
- What characteristics do good written and oral presentations have in common?
- How might visual aids detract from or add to an oral presentation?

LEARNING OBJECTIVES

By the end of this chapter, you should be able to:

- Identify key characteristics of effective oral communication
- Recognize the importance of pacing in oral communication and understand methods by which to measure your own effective pace
- Determine the level of formality necessary for a given presentation based on the audience, venue, and type of presentation you are giving
- Summarize critical considerations associated with the different roles you may have in video conferences
- Compare and contrast best practices for visual accompaniments to oral presentations with those same presentations if the visuals are designed to stand alone

INTRODUCTION

In the previous chapter, our expectation regarding written communication was that you had already taken or were going to take at least one or more college-level courses on writing. For some of you, the same may be true for oral communication, although we suspect that the number of students for whom this is true is much smaller than it is for written communication. Oral communication is one of those areas that you will likely be expected to be proficient in over the course of your professional career, yet it is quite possible that you will never have a formal course covering this topic. Again, we do not suggest this chapter by itself can cover the myriad of material necessary to appropriately master this topic area. Instead, we hope this chapter emphasizes things you already know about effective communication (learned either inside or outside the classroom), provides additional considerations and methodologies to improve your communication skills, and gives you insight into the areas of oral communication you can improve to be a more effective presenter in your future classes.

Although you may have never had a formal class on public speaking and oral communication, certainly many of the things we described in the previous chapter related to effective written communication are equally pertinent to effective oral communication. Concepts such as having a clear, concise message; treating the entirety of your communication as a story; and using language that can be easily understood by your audience equally apply to both oral and written communication. Over the course of this chapter, we will dive a bit deeper into some of these topics as they pertain to oral communication, as well as cover some topics that have emerged over the last few years due to how many of us are now communicating across different rooms, cities, and countries. We will also be talking about using visual aids to enhance your oral communication, which will extend some of the discussion we had in our previous chapter about the use of visuals (primarily charts and graphs) in written communication. At the end of this chapter, we will briefly touch upon how these suggestions for communication within your profession will likely influence your personal communication and vice versa.

Written communication concepts such as a clear, concise message, treating the communication as a story, and using language that can be easily understood by your audience apply to oral communication as well.

SPEED, TONE, AND VOLUME

If you have ever played a musical instrument (yes, we'll even include the recorder as a musical instrument), you know that the melody of a particular song is a function of the pitch of the various notes in the composition, the rhythm of those notes, the dynamics associated with playing those notes and the tempo at which the notes are played. Translating that to speech, we have speed (tempo), tone (pitch and rhythm), and volume (dynamics). Just like an interesting melody will often incorporate all of these in different ways, so too will an interesting speech.

Much of what we will cover in this chapter focuses on best practices we have come to expect in our own classrooms and are areas where small changes can lead to a significantly improved presentation. One of the first areas we find that most students can improve their oral communication is simply in the mechanics of how they communicate to their audience. As we will discuss in the next section, one of the things you will need to consider is the level of formality necessary for your given oral presentation. However, regardless of the formality level, one should not view that formalism as a requirement to be less dynamic in their speaking mechanics than they would be in a group of friends or family. On the contrary, **we would suggest your own personality should come through in your public speaking, and one way to make sure this happens is to be thoughtful in the speed, tone, and volume of your communication.**

Many of the students we talk to regarding oral communication admit to the fact that they have a strong fear of speaking in public. If this sounds like you, recognize that glossophobia (the clinical name for fear of public speaking) is the most common phobia people have (73% of the population), exceeding the fear of death or of spiders[1] (and certainly the fear of death by spiders). As such, when they get up in front of an audience, they speak very differently than when they are talking with their friends and family or are even engaging with individuals in a less formal setting or a one-on-one situation. Nerves can lead to multiple problematic speech characteristics, several of which we will address in this section.

> **Engineering in Action**
>
> Break off into groups of three or four. Discuss approaches that can be taken to reduce the anxiety associated with speaking in public. These can be something you have tried or something that you have heard of or read about as ways to be more comfortable speaking in public.

---

1 https://nationalsocialanxietycenter.com/2017/02/20/public-speaking-and-fear-of-brain-freezes/.

Speed

If you have ever been required to give a presentation with a specific time requirement, you might have found that even though you practiced and rehearsed that presentation over and over until your timing was impeccable, on the day of the actual presentation, your timing was much faster than what you expected based on your rehearsals. Speaking faster is a common result of being nervous about a presentation. Remember, conversational speech rates (120–150 words per minute) are much slower than common reading rates (200–300 words per minute or more). Also recognize that public speakers may speak at speeds faster or even slower than conversational speech rates to get a specific point across (watching or listening to recordings of public speakers online can give you a good sense of the range of speech speeds that can be effectively used when delivering a presentation[2]). **By speeding up, you can convey a sense of excitement to your audience, while slowing down might suggest you are trying to make sure they fully understand a critical part of your presentation.** Typically when nerves set in, everything is just one speed: fast. For some presenters, that speed also comes at the cost of enunciation and articulation, that is, the clarity and physical action of forming and pronouncing your words. When this occurs, words can lose their "crispness," and the speaker may sound as if they are mumbling (not in terms of volume, but in terms of word formation, we will talk about volume a bit later).

Greater speed in speech can often negatively impact enunciation and articulation that is the clarity and physical action of forming and pronouncing your words.

If you have ever heard a car advertisement on the radio, the announcer will often speed through the legal description of the terms and conditions of the financing associated with the new car at the end of the commercial.[3] Although the announcers are able to speak at these significant speeds while maintaining appropriate enunciation and articulation, they are speaking at such a speed that you need to really listen attentively to understand what they are saying. Additionally, the speed becomes monotonous and monotone, making that portion of the ad much less engaging than the rest of the commercial. While this is an extreme example of speed, and the lack of engagement for the legal portion of the ad is done on purpose, it provides an example of what can happen if you talk too fast without varying your speed, tone, and volume. **For any oral presentation you give, you want to make sure to stay at a pace that is comfortable for you and your audience.**

---

2 https://improvepodcast.com/words-per-minute/.
3 https://www.youtube.com/watch?v=9s5W78LMlM4.

So, what can you do to ensure you are not talking too fast, particularly in those cases when you have already rehearsed your speech multiple times and find yourself talking too fast during the actual presentation? One thing to try, particularly when you are practicing a presentation, is to record yourself and watch that recording afterward. Try to remember how that pace felt to you as a speaker, and then critique that pace as a listener. For many of us, the speed that feels good when we are talking feels too fast when we are the listener. If this sounds like you, feeling as if you are going too slow in your public presentations might very well mean that you are actually talking at a reasonable pace. Understand that your audience may not be as familiar with the information you are trying to convey as you are, and as such they will need to not only hear what you are saying but will also be expending additional energy attempting to understand or process your information.

Another way to slow down your rate of speech is to try to create purposeful pauses in your presentation. This could be something as simple as making sure you are pausing (and breathing) after so many words or being purposeful in adding a pause after each point in a list. **Being cognizant of these opportunities for silence during an oral presentation is just as important as white space is on a paper.** Talking nonstop for 10 or even 5 minutes is very taxing to a listener, and the longer you as a speaker go without a pause (or breath), the harder it is for the listener to continue to concentrate on what you are saying.

Make sure the amount of content you want to present is appropriate for the presentation time.

One additional thing to consider when it comes to speed, although this suggestion could easily be placed in several other sections in this chapter, is making sure the amount of content you want to present is appropriate for the amount of time you have for your presentation. How many times have you come to class and a feeling of dread comes over you when your instructor opens their PowerPoint slides with thirty minutes remaining in the lesson and you see they have eighty slides to cover? Even if several of those slides are "reveals" or graphic changes (versus using transitions within the PowerPoint program), the sheer volume of slides suggests there might be a disconnect between how much the instructor is attempting to cover versus how much time they have to cover it. Being thoughtful and purposeful in the development of your visual aids (see later in this chapter) and making sure your story is focused and to the point (similar to what we described in Chapter 5) can also reduce the subconscious need to go faster to cram everything in when you have too much material to cover for a given duration of time.

Tone

As we described at the beginning of this section, we can relate speaking tone to the pitch and rhythm of a melody. Although this is not a perfect analogy (often tone will also encompass volume and speed in a way that is difficult to separate), thinking of tone as the arrangement of the melody should help you understand what we are talking about. When we use the word tone, we are describing the changes in the dynamics and pitch of your speech patterns to emphasize certain words, phrases or ideas. This may also result in a change of volume and perhaps even the use of non-verbal communication such as gestures and posture. One easy way to really understand this concept is to go watch any stand-up comedian and try to separate the words they say from the way they say it. The delivery of the jokes, which encompasses volume, speed, and *tone* is exactly what we are talking about. Tone can express emotion, and in terms of an oral presentation, those emotions could range from excitement, to importance, to perhaps even disappointment, depending on what the situation calls for. **In many ways, the tone of your delivery can be as important as the words you say when it comes to creating a memorable presentation.**

If you have ever seen *Ferris Bueller's Day Off* (or even if you have only seen the clip of what we're describing on YouTube), Ben Stein's example of a monotone high school teacher (Bueller …) is something we want to avoid when giving a public speech. Obviously, if someone is very nervous, their tepid or timid delivery might result in a monotone, where there is no inflection or change in their speech tone. Just like the example of Ferris Bueller, or our example of the lack of inflection with a speed talker, it makes the listener work harder to remain engaged in the presentation. One other issue we sometimes see with student presentations is the thinking that formality should repress or remove the use of inflection and changes of tone in a presentation. While formalism might impact the types of tones used during your speech, by no means does it mean you should completely forgo the use of changes in intonation. If you were to make an analogy between oral presentations and art, you could argue that tone is the pallet of colors an artist uses in her/his/their paintings. When comparing the tone(s) you might use in everyday conversation to those in a formal presentation, you can think of those as two different pallets, where some of the colors may overlap, and others might be completely separate. Even if you look at great artists such as Pablo Picasso and analyze the pallets that he used in his "monochromatic" phases such as his Blue Period, his paintings during that era contained many different shades of blue and blue green, in addition to the occasional use of other colors. Very few people communicate in their day-to-day lives in a monotone fashion; when speaking during a presentation, this should also be avoided.

While the types of tones used during a speech might be impacted by the level of formality, you should not completely forgo the use of changes in intonation.

Volume

The third main topic in this section is the appropriate use of volume, or dynamics, in your speech. Similar to the previous topics, where we encourage you to vary the speed and tone of your speech to highlight or emphasize various aspects of your presentation, so should you (and perhaps somewhat by extension) also attempt to vary the volume of your speech patterns. For many speakers, when they get nervous, the variability of their volume goes away, and that can lead to someone being consistently too soft (most often) or consistently too loud (less often, but just as problematic). In either case, **speaking too loudly or too softly can make listening and comprehension much more difficult**, as your audience may be straining to hear what you are saying (too soft), or trying to find a way to limit your volume (too loud, particularly if you are amplified by a microphone or perhaps in an online environment [see Video Conferencing for additional information]). Additionally, changes in dynamics can help emphasize your point and add variety to your presentation, much in the same way speed and tone can enhance your presentation.

There are multiple factors that can influence whether a person is naturally loud or soft, or perhaps somewhere in the middle.[4] Regardless of what type of talker you are, most people will naturally change the volume of their voice over the course of an informal conversation, perhaps becoming louder when delivering the punchline to a joke or the reveal of a scary story or becoming softer when talking about sensitive information. These dynamic changes are audio cues that are well known to a listener, helping them better understand the nature of the information that is being provided. Taking advantage of these cues and using a variety of volumes as you speak can help draw people in and keep them focused on what you are saying throughout your presentation.

As we will describe in the next section (Level of Formality), a formal presentation does not need to be a somber event, using your "library" voice. While the level of formality can help determine the level of detail and language expected within a presentation, formal does not equal stuffy, non-dynamic or boring! By using the same speaking tools you likely already use in your everyday conversations (speed, tone and volume), your public presentations will become more dynamic, engaging and memorable. Formality should never equate to lack of personality or personability; these are attributes that will be unique to you and should come through in your presentations.

> **Engineering in Action**
>
> Break off into groups of three or four. Discuss within your group what you consider to be effective oral communication characteristics (these can and should go beyond speed, tone, and volume).

---

[4] https://www.nbcnews.com/healthmain/loud-talkers-why-do-some-voices-seem-be-set-top-437046/.

LEVEL OF FORMALITY

One important factor to understand when preparing an oral presentation is who your audience is, and by extension, what the level of formality is regarding the venue and event at which you will be presenting. When we talk about formality, we are focusing on a few different things that should be considered prior to your presentation planning and, ultimately, your presentation itself. Although each of these we will list as separate considerations, they are very often synergistic and, as such, cannot be considered independent of one another. When combined, these three considerations will also influence other aspects of your presentation that we will not cover in this chapter, such as dress considerations. In particular, we consider the following:

1. The **type of audience** and their prior knowledge of the topic you will be discussing/presenting. Also included in this is your familiarity with the audience
2. The **type and size of the venue** in which you will be presenting
3. The **type of oral presentation** you will be giving (speech, pitch, testimony, discussion, etc.)

Within these considerations, we quickly see how different situations may call for different levels of formality. Preparing to have a conversation via a podcast that will be consumed primarily by novices of a particular field of study will look and sound very different compared to giving a talk at a prestigious conference or giving testimony at a public hearing. Compare Neil deGrasse Tyson and his demeanor, language, and tone when he gives testimony to Congress[5] versus when he talks during his StarTalk podcast.[6] You can hear distinct differences in not only the language used but also the dynamics, tenor, and speed of his speech. You will also notice that his dress differs significantly in the two presentations. Although both talks include the full range of speech changes we describe in the previous section, we see that the formal presentation is different than that used in his podcast due to where it was held and who he was addressing. In both cases, Dr. deGrasse Tyson is an engaging, knowledgeable speaker; however, he chooses to use different tones, volumes, and speeds in each situation to better match the level of formality of his surroundings.

One of the first rules of developing an engaging presentation is to know your audience. Who will be present at your talk? What will be their level of understanding or knowledge regarding the topic you will be covering? Will you need to simplify your talk, providing a significant amount of background information just to ensure your audience will be able to follow the rest of your presentation, or is the audience sufficiently educated regarding the topic that you can jump directly into the major findings associated with your presentation with just a minimal amount of background (more as a refresher than attempting to bring your audience up to speed)? What information is presented, the language you use during the presentation, the

---

[5] https://www.youtube.com/watch?v=rmKlA_UnX8c.
[6] https://www.youtube.com/watch?v=aYBEcFvI8o8.

amount of information contained within your visual props, and more will all be dictated by the type of audience to which you are presenting. What you say at a community meeting that is open to the public versus what you talk about in a design meeting with the rest of your design team that has been working on the same project for months or years may be very different, and very different language may be used in those two situations to describe the same concept. While this might be somewhat peripheral to the level of formality, it does play a large role in how technical your talk might be and, as such, may change the formality of the language you use to ensure your audience remains engaged.

The audience of your presentation dictates the information presented, the language used during the presentation, the amount of information presented, and more.

The venue itself will also influence how formal your talk will be and may require very different approaches to how you give your presentation. In a large auditorium or venue, it might be quite difficult to see your body language compared to a more intimate conversation amongst a handful of individuals. Amplification systems such as microphones are another consideration that may limit your ability to move freely throughout the venue and may require that you remain stationary at a podium (particularly if the microphone is attached to the podium). Additionally, your use of visual aids may be limited to only using slides or pictures versus potentially using physical props such as prototypes or models. In each of these circumstances, the nature of your presentation will be altered by these limitations (although more correctly, we would define these as specifications of the venue, as the word limitation suggests one type of audience, location or presentation is superior to another). These differences can also potentially affect the type of presentation that is ultimately made (for example, a large venue might make having a one-on-one conversation with a single member of the audience more difficult). While this may change your approach to the presentation, it does not limit your ability to be dynamic in terms of your voice. In circumstances where your use of non-verbal communication is compromised (body language and/or gestures, facial expression, etc.), you can still convey your message through the appropriate use of language and changes in the inflection, timbre, and speed of that language. Again, formal does not mean boring, wooden, or lifeless!

The third consideration when thinking about formality is the type of presentation you have been asked to do. Again, this may very well be a result of the audience and size of the venue but can also be independent of these considerations. Leading a discussion amongst co-workers or colleagues is very different from testifying before a judge or governmental entity. Each has a distinct level of formality and a separate type of preparation. In your discussion, you may be very informal, telling colleagues to interrupt at any time and deferring to their own expertise as they explain certain results or conclusions. This will likely be different compared to testifying before a county board, where you are being looked upon as the technical expert in your field, and it is your responsibility to ensure that the novices in the room have enough information to make an appropriate decision that could have long-term implications for your community. Again, it is difficult for us to separate these three considerations, and as such, it is likely some combination of all of them will affect the formality of your presentation.

VIDEO CONFERENCING

One thing that has come to the forefront during the COVID-19 pandemic that was less of a consideration amongst universities and corporations prior to 2020 is the use of video conferencing to hold meetings and conferences. Due to travel and room occupancy restrictions resulting from the global pandemic, everything from five-person discussions to fifty-person meetings to five-thousand-person conferences have been put online, resulting in new considerations in regard to presentation etiquette (both from the perspective of the speaker as well as the audience), formality (hey, no one can see me from the waist down, so pajama bottoms it is), and more. While for some, meetings and conferences have returned to a more familiar in-person format for presentations, it has become readily apparent that video conferencing will continue to be extensively used, in both academia as well as industry for the foreseeable future. As such, let us look at a few things to consider, particularly when it comes to presenting within a video conference, that are different than the scenarios we have previously described.

> **Engineering in Action**
>
> Take two minutes as an individual to list some of the best practices and most annoying habits you have encountered during virtual presentations. Then discuss these with your team and compile a list of the top best practices as well as the top things to avoid.

Cameras (Where to Look), Lighting, and Sound

One thing to think about when it comes to video conferencing is what type of device you are going to use for that conference and where you are going to use it. An issue that comes up in many meetings is the location of the speaker's camera relative to where they are and where their device is located relative to the lighting sources in the room. On many computers, whether it is built into the system or a separate entity, the camera is located at the top of the computer screen. Depending on your desktop setup, that may be higher or lower than where your gaze naturally occurs when attempting to look at your audience. As such, unless you are making a conscious effort to look directly into the camera, you will not appear to be making eye contact with your audience, even though you may be looking directly at them on your computer screen. Although it may feel awkward to not be establishing eye contact on your end, by looking into the camera, you are establishing eye contact on the listeners' end, which is the more important rapport to establish.

Looking at the camera establishes eye contact from the listeners' perspective, which improves rapport.

In addition to making sure you are looking directly into the camera whenever possible, try to create an environment where the lighting is appropriate for your presentation. Like taking a good picture, make sure you have a good amount of diffuse lighting in front of you and try to avoid having a significant amount of lighting behind you. This might mean working in an area where you can be facing (or facing at an angle) a window or having desk lamps or other forms of incandescent lighting that you can use to front-light your work area. One thing to be cautious about when mixing natural light with incandescent lighting is attempting to match the temperature (hue) of the different types of lights; a temperature mismatch can cause orange or blue tones to be cast, which is often far from flattering.

The third thing to consider when using video conferencing technologies is your sound. Particularly if you are working in an environment where sounds from your surroundings can be distracting, or perhaps even picked up using an open microphone, **the use of headphones and a microphone are highly encouraged**. This can be something as simple as the set of headphones that came with your cell phone or something as complex as a studio monitor with a separate high-end dynamic or condenser XLR microphone. Either will likely be better than the built-in microphone on your computer, and by keeping the microphone closer to your face (and further away from your computer), it is less likely to pick up the ambient noises of your device (keyboard, mouse buttons, fan, etc.). As we discussed in the previous section, the dynamics of your vocal presentation play an important role in telling your story to your audience and using these extra tools will make sure those nuances come through without hearing the rest of your surroundings. Additionally, by using a headset, you reduce the possibility that your microphone will pick up the sounds coming out of your speakers, which leads to the echo feedback you have probably experienced when a colleague who is not using headphones has her/his/their microphone unmuted when someone else is talking. Even though many video conferencing software packages attempt to minimize this feedback, it does happen and can be very distracting during a presentation.

Etiquette

Video conferencing etiquette is something that continues to evolve and is something that, is not a one-size-fits-all approach. There are nuances to what we will describe in this section, and we recognize that everyone's situation might warrant a different approach for presentation purposes.

In the end, the goal of such etiquette is to demonstrate your attentiveness to the presentation and/or audience, to engage with the other members of the video conference, and in the process, minimize any technology distractions.

The goal of video etiquette is to demonstrate your attentiveness to the presentation/presenter/audience and engage with other members of the video conference, all while minimizing technology distractions.

As we suggested in the previous section, particularly when giving a presentation, having quality lighting for your camera, looking into the camera when speaking, using an external microphone to enhance your sound, and using headphones should you need to hear what is going on in the conference while your microphone is unmuted are critical to avoid having your messaging clouded by technology limitations. Similarly, if you are listening to a presentation or participating in an activity where there is a back-and-forth conversation between you and one or more additional participants, you want to demonstrate your engagement with the speaker as appropriate or recommended by the host of the conference. Many times, particularly for large conferences, audience members will be muted and will either be unmuted by the host or the audience member will be given the opportunity to unmute themselves when it is time for them to speak. **One of the tricks here is recognizing when you are muted and when you are not muted; both situations can present problems if you are unaware of your status.** If you are involved in a video conference where you are not muted, making sure your audio is not causing a distraction is vitally important. Most people will keep themselves muted specifically for this reason and only unmute when they have something to add to the conference. Of course, doing so can cause you to forget that you are muted, and when called upon, you may end up talking without anyone hearing you, so just remain aware of your status and proceed appropriately.

Another etiquette question that depends on the circumstance is whether or not it is necessary to keep your video on during a presentation as an audience member. There may be no "right" answer other than to follow the lead or suggestion of the host or other members of the video conference. For some individuals, the speed of their internet connection may necessitate the turning off of their video in order to ensure they have a stable connection to the conference. Others may feel self-conscious about their appearance or surroundings or may be in a situation where it could be distracting to have their video on (for example, if someone is working from home with a young child who is in the same room). As a presenter, it is difficult to give a presentation without seeing the faces of your audience. Just as presenters provide non-verbal cues to their audience during their presentations, audience members provide non-verbal cues to the presenter that are valuable in steering the direction of the presentation. Looks of confusion might suggest certain topics need further clarification, while looks of boredom could suggest to the presenter to move along in their presentation or find another avenue to re-engage the audience. Additionally, nods of understanding, smiles, or even laughter (particularly if you intended to be funny) can energize and encourage the speaker, showing her/him/them that they have hit the mark in engaging with the audience. All of these non-verbal cues are lost without the audience members having their video camera on. **Our current recommendation on this topic is to turn the camera on if your situation allows for it** and provide the speaker with the same level of non-verbal feedback you would normally give if you were attending the presentation in person.

One additional etiquette piece to consider is thinking about adding in a delay that feels longer than necessary when pausing for questions or when waiting for the next speaker to start speaking. While some of that delay is due to people trying to unmute themselves to begin speaking, some of the delay is people trying to make sure they are not speaking over someone else. I'm sure we have all been in conferences where two or more people attempt to speak simultaneously, both stop, then both start again almost at the same time. Because we lack some of the non-verbal cues we are used to seeing when meeting in person, there can be pauses in the flow of a conversation or delivery of information that are different from an in-person event. Should you have these interactive opportunities during your presentation (asking audience members for their feedback, providing an opportunity for the audience to ask questions, etc.), building some extra time into your pauses prior to continuing with your presentation is warranted and will likely allow for a more engaged audience. Additionally, suggesting that audience members submit their questions or comments via chat (provided you can stay on top of that medium while presenting or have a second person such as a moderator following the chat) is a way to keep your audience engaged and may also allow for voices that would otherwise not participate to become better engaged in your presentation.

Building extra time into your pauses when presenting virtually can increase audience engagement.

POSTERS AND SLIDE PRESENTATIONS

As we suggested earlier in the chapter, your visual props, such as posters and slide presentations, will depend on the type of presentation you are giving, your audience, and location. A slide presentation for a course lecture will likely look very different from a poster presentation for a research conference in terms of content, but both will follow some of the same rules regarding formatting, color, etc. We touched upon aspects of formatting documents back in Chapter 3 when it came to resumes, and some of the information there regarding font and color will continue to apply here. As this chapter is focused on oral presentations, **our goal is to touch upon aspects of the poster and slide presentation that directly affect an oral presentation.**

Content

One of the first things to consider when determining the layout of a poster or a slide is what content you want to share with your audience. The type of presentation you are giving, as well as the makeup of your audience, will likely influence the content you want to include in your visual props. For example, if you are giving a classroom lecture, your slides might include additional information such as specific equations or definitions you want your classmates to remember or learn, whereas a presentation containing similar material given as part of a presentation for a design review might contain less information directly on the slide, as there may be an accompanying report that each member of the team will have that contains that pertinent information. What this comes down to is what is it that you want your visual props to do? Specifically, visuals can be broken down into two distinct categories:

1. **The visual needs to "stand alone,"** that is, it can or will be viewed or referred to independently of the initial presentation. These types of visuals can include classroom slides that are intended to be used for studying, poster presentations at a conference that will be viewed independently of the author's presentation, or a rendering of a product or building that will be on display for review and comment by the public. In each of these circumstances (these are only a few of many examples that could be given), the visual will be used beyond simply being a prop meant to enhance the oral presentation. As such, these are likely to contain a more significant amount of information such that they hold value or tell a story outside the context of the original presentation.

2. **The visual is used to enhance or reinforce the material being discussed within an oral presentation** and will primarily be used only to accompany that presentation. These might be PowerPoint slides containing data from a study or images of an upcoming product or building. Independent of the oral presentation, these types of visuals do not provide enough information for an independent viewer to fully understand the context of the images or information, but in conjunction with the oral presentation, these visuals provide a more complete picture of the information being described by the speaker.

The type of visual you are producing will play a large role in determining what kind of information you include in your slides or poster. While both will be similar when it comes to slide composition (see below), the content of each will be specific to your application and need. As such, it would be very difficult to provide overarching guidance regarding the type of information that should be included in a slide deck or poster that accompanies an oral presentation. The one thing we will stress when putting together such a presentation is to be conscious of the story you want to tell and how the visual will aid in the telling of that story.

Visuals that are designed to stand alone often contain a greater amount of information than those visuals that are designed to provide a more complete picture as part of an oral presentation.

As we suggested in the previous section, one mistake people commonly make when using slides as part of their oral presentation is they use too many slides, thinking that if the information is presented in a visual manner, it will allow them to cover the material faster (somewhat akin to the idea that a picture is worth one thousand words). Moving too quickly through a slide presentation is just as off-putting to an audience as talking too fast, so it is best to be conservative in the number of slides you use. There are plenty of "rules" and "suggestions" that some students are looking for in terms of how many slides their presentation should contain. Many of these "rules" will state anywhere from one slide for every thirty seconds up to one slide for every three minutes. Our philosophy has always been to use the number of slides you need to get your point or story across to your audience. Worrying about hitting a specific number of slides might be required for certain presentations, and in those cases, you should adhere to those specifications, but in general, we have seen very effective presentations where the number of slides used was on the higher side of the range described, and others that were at the lower end of that range. **Focus on the content you need to be successful in your messaging.**

Color and Composition

As we previously described in Chapter 3 when we discussed résumés, color and composition can play an important role in presenting your information in a professional, legible fashion. Similar to that section, your choice of layout, font size, font type, and the use of bolded, underlined and/or italicized text can easily emphasize an important concept, but it can also distract from the intent of your presentation. Additionally, slide shows offer the opportunity to include additional visual emphases, such as reveals, dissolves, wipes, and other transitions between or within your slide. Furthermore, this type of presentation allows for the inclusion of video or other audio-visual effects that can be powerful but can just as easily be overpowering. The choices you make in the development of your presentation will likely be influenced by the format and formality of your presentation, including such considerations as the type and size of your audience and venue as well as the type and length of your presentation. Again, the combination of factors is such that it becomes difficult to provide specific guidance; as such, we will provide the following list of things to consider when developing your visual aids.

1. **Consider the venue and size of your audience when determining font type and size.** Make sure your font is large enough that anyone in the room in which you will be speaking can read it. In the case of a poster, recognize that your audience will likely be anywhere from 3–6 feet away from your document (and perhaps as far as 10 feet away), so you want to choose a font size and type that can easily be read from the maximum distance. Using one or at most two traditional serif or sans-serif fonts without drop-shadows or any other type of font effect is a good practice. For slides, using font sizes less than 24 point is often discouraged, although this will somewhat depend on a number of factors, including what font type you have used in your presentation, the size of the room you are presenting in (or if you are presenting virtually) and the size of the screen your slides are being projected onto, to name a few.

2. **Composition is very important to your messaging.** Just as we described for your résumé, the composition of your visual aids is very important. In addition to choosing an appropriate font and size, make sure the layout of your document includes an appropriate amount of text, white space, and images (or other visual aids such as movies, graphics, etc.) and that these elements are positioned in such a way that one can easily understand the order in which they occur. Recognize that for most western readers, their eyes will scan a document from left to right, then top to bottom, just as you are reading this book. As such, they typically expect the layout of a document such as a poster to follow this orientation. Similarly, for slides, viewers will likely be drawn toward the top of the document, and as such, your layout should account for these natural tendencies.

3. **Colors are important.** As we also discussed in our section on résumés, the choice to use (or not use) color is an important one, and when using color, certain combinations are better than others for purposes of clarity and contrast. For visual aids for oral presentations, these rules still apply, and our understanding of appropriate choices of color combinations will need to be expanded to not only include printed material such as posters but also projected materials such as computer graphics or slides. Looking back at the Résumé section in Chapter 3, we found for both projected as well as printed materials, dark text (black, blue, red, brown, and violet) on white or light backgrounds were the most legible for the average reader.[7,8,9] One area that these studies did not address was considerations for audience members that might have impaired or compromised color vision, such as those individuals who are color blind (with the most common combination being red-green color blindness which affects up to 8% of all males and 0.5% of all females of Northern European descent[10]) or the elderly.[11] These limitations should be considered when looking at various slide layouts, regardless of what the software manufacturers present as a standard template. Furthermore, while posters are often composed using computer software, and your resulting layout may look really good on an LCD screen, ensuring that the poster printer you use will print an even, flawless dark-colored background without any bleeding into the light-colored text is not as straightforward as it sounds. While we are not saying that color should not exist in your visual materials, understanding the potential limitations associated with your audience's ability to see said materials or the ability to translate your digital masterpiece onto a printed piece of paper should be accounted for in your decision to add (or not to add) color to your

---

7   Iztok Humar, Mirko Gradisar, Tomaž Turk, and Jure Erjavec, "The impact of color combinations on the legibility of text presented on LCDs," *Applied Ergonomics*.

8   T. Nilsson, "Legibility of Colored Print," *International Encyclopedia of Ergonomics and Human Factors*.

9   Greco, Massimo et al., "On the portability of computer-generated presentations: the effect of text-background color combinations on text legibility," *Human Factors*.

10 Bang Wong, "Points of view: Color blindness," *Nature Methods*.

11 Keiko Ishihara, Shigekazu Ishihara, Mitsuo Nagamachi, Sugaru Hiramatsu, and Hirokazu Osaki, "Age-related decline in color perception and difficulties with daily activities," *International Journal of Industrial Ergonomics*.

presentation. A more sparing use of color might actually be more effective than its liberal use as it may bring greater attention to the item(s) that include color. As highlighted in our next topic, sometimes less can actually be more.

4. **Less is more.** Particularly when it comes to a slide, too little text is generally going to be better or at least easier to fix than too much text. While it may be tempting to include as much text as possible on a slide to use as a prompt or as talking points for your presentation, reading your speech from your slides or poster will not lead to a memorable presentation (at least not in the positive sense). Furthermore, too many visuals on a single slide, whether that is text or other forms of visual information, will quickly lead to audience fatigue, resulting in both a less engaged audience and a less effective visual effect. As we suggested earlier in this section, you might find yourself using more text than you normally would if your visual aids need to "stand alone" (as would be the case if you were creating slides for a class presentation, for example), but even in that case limiting the number of words (or more importantly the number of ideas) on each slide will remain an important consideration.

Maintaining Rapport with Your Audience

One mistake people often make when using a visual prop is that their attention immediately switches to the prop rather than maintaining their connection with the audience. This can be something as obvious as turning your back to the audience to point out a feature of the slide, to something less obvious (but equally distracting), such as hovering over or continuously circling a key point on a slide with a laser pointer. In both cases, the presenter's attention is now focused on the slide or poster, and they are no longer focusing on the audience. For presentations in larger spaces, this can be compounded by the fact that you might have multiple screens that contain your visual prop. Focusing on only one of those screens immediately alienates the remainder of the audience that is not near the screen you have chosen to address during your talk. In general, much of the need to point out specific aspects of a visual can be eliminated by having a thoughtful, well-composed visual aid that does not contain too much information and only one (or even only a part of one) idea. By doing so, your visual will satisfy the needs of the audience and complement your presentation without requiring you to point something out or telling the audience where to look. If you have ever been in a presentation where the speaker apologizes for the slide being too busy or the font being too small, they obviously did not take into consideration the audience's needs when crafting their presentation and visual aids.

Well-composed visual aids that contain one (or part of one) idea reduce the need to point out specific aspects of a visual, satisfy audience needs, and complement your presentation.

For the most part, what we described in the previous paragraph are potential pitfalls primarily associated with the use of slides during a presentation. Posters, on the other hand, by their nature, cannot be limited to a single idea, and as such, during a presentation you may need to be more direct in guiding your audience through the document or pointing out visuals within the poster itself that reinforce or complement your presentation. However, this is to be expected of a poster presentation and thus will not come across as being a distraction to your audience. What you do not want to do, however, is to focus your attention on the poster for too long. It is okay to bring an audience member over to point them to a specific visual that enhances your presentation, but your focus should remain on the audience as a whole. One tricky part of a poster presentation is what to do should individuals arrive (or leave) as you are giving your presentation to another group of individuals. **Attempting to engage the entire audience may prove difficult, but at a minimum, acknowledging the entirety of the audience either through verbal recognition or via eye contact or other non-verbal cues is encouraged.**

CHAPTER SUMMARY

This chapter focuses on effective approaches to oral communication, including the integration of visual effects into various oral communication situations. Within this chapter, the following topics were covered:

- The impact of speed, tone, and volume of your voice on oral presentations
- The way in which the type of presentation, the type and size of the presentation venue, and the type and background of the presentation audience influences the formality of your presentation
- Considerations for video presentations, both from the aspect of a presenter as well as an audience member
- Aspects of poster and slide presentations, such as content, color, composition, and the manner by which you ask the audience to engage with your visual aids that influence effective oral presentations.

END OF CHAPTER ACTIVITIES

Individual Activities

1. Find a paragraph or set of lines from one of your favorite books, plays, or TV shows and record yourself reading these lines. Now find an abstract from a technical article or the introduction to one of your class activities and record yourself reading these lines as well. Compare the recordings considering the following:
    a. What is your level of formality for each talk? Are these similar or different?
    b. How does your speed, tone, and volume vary for each talk?

    c. As you watch your recordings, which do you consider to be an engaging presentation?
    d. If one recording was significantly less engaging, how could you adjust it to be a more engaging experience for the audience? Try re-recording that talk to integrate these thoughts and re-watch to assess your success.
2. Consider a recent (or upcoming) presentation of yours. How did/will the audience, venue, and type of presentation influence five key features of the presentation. Be explicit regarding how these features result in a specific approach, design, consideration in the presentation.
3. Identify three locations to which you currently have access that you consider most effective for your participation in a video conference. What qualities do these sites possess that make them the most effective? Is one significantly better than the others? Why or why not? Would all work equally effectively for when you are an audience member in a webinar, a presenter for a virtual conference or class, or a candidate in a virtual job interview? What are the potential pitfalls of each location?
4. Create a two- or three-slide presentation on a recent concept from one of your classes. Make sure it contains at least one picture, figure, or graph. First, create this presentation as if it was going to be used as a "stand-alone" summary of the material. Then revise it as if it was going to augment an oral presentation on the topic. Compare the two presentations and highlight three key similarities and three distinct differences between the two.

Team Activities

1. As individuals, select a short speech and practice it before coming to class or the group meeting. You do not have to create this from scratch. Pick something you have heard before but select something that resonates with you. Bring a copy of the speech to your group meeting. Before you present your speech, switch speeches with someone else in the group.
    a. Have each individual read one of the other group members' speeches out loud.
    b. After everyone in the group has read someone else's speech out loud, return the speeches to the original "owners" and have them read their speech out loud.
    c. As a group, discuss the following:
        i. What similarities and differences did you witness between different individuals giving the same speech?
        ii. What similarities and differences did you witness between different speeches from the same individual?
        iii. What were the top two or three strengths of each individual's presentations?
        iv. What were the top one or two items to improve for each individual's presentations?
2. As a group, discuss which type of presentation, venue, or audience you have presentation experience with to date. Which do you anticipate having to present to in your academic career? After graduation? Which do you consider the easiest (i.e., least stressful) and why? Which might be the most stressful? What suggestions does the group have for reducing stress in those most stressful situations?

3. Schedule a virtual team meeting. As individuals, make sure you are at a location you would consider to be appropriate if you were conducting a virtual interview. As a group, provide feedback to each member on the location each of you selected (i.e., background, lighting, sound, etc.). Take turns serving as the meeting "host," asking questions, facilitating discussions, etc. What did the group learn from this experience?
4. As a group, select a presentation to analyze. It can be one created by a group member, one from class or a workshop, or something found online. Have each individual view the presentation prior to the group meeting. In your meeting, discuss the strengths and areas for improvement, focusing on concepts discussed in this chapter.

CHAPTER REFERENCES

Greco, Massimo, et al. 2008. "On the Portability of Computer-Generated Presentations: The Effect of Text-Background Color Combinations on Text Legibility." *Human Factors* 50, no. 5 (October): 821–33. doi:10.1518/001872008X354156.

Humar, Iztok, Mirko Gradisar, Tomaž Turk, and Jure Erjavec. 2014. "The impact of color combinations on the legibility of text presented on LCDs." *Applied Ergonomics* 45, no. 6 (November): 1510–17. https://www.sciencedirect.com/science/article/pii/S0003687014000696.

Ishihara, Keiko, Shigekazu Ishihara, Mitsuo Nagamachi, Sugaru Hiramatsu, and Hirokazu Osaki. 2001. "Age-related decline in color perception and difficulties with daily activities-measurement, questionnaire, optical and computer-graphics simulation studies." *International Journal of Industrial Ergonomics* 28, no. 3–4 (September): 153–63.

Nilsson, T. "Legibility of Colored Print." 2006. *International Encyclopedia of Ergonomics and Human Factors*, Second Edition. 3 Volume Set. CRC Press. http://dx.doi.org/10.1201/9780849375477.

Wong, Bang. 2011. "Points of view: Color blindness." *Nature Methods* 8, no. 6 (June): 441.

CONCLUSION

THE END (AND BEGINNING) OF YOUR JOURNEY

As we have put together this textbook, we have given careful thought as to how it might be used in the classroom and who might be included in the audience that reads it. We recognize and suspect that as part of a course, specific chapters might be assigned in support of that course's goals or outcomes, and as such, some (most?) readers will not necessarily read this text cover to cover. That being said, we did want to create an additional section at the end of the book that ties things together and perhaps broadens the impact of the topics we have covered across the individual chapters. As you may already notice from the initial structure, this last section is less about providing new information or teaching elements and more of a reflective piece, providing you with some things to perhaps take away from the book and think about moving forward.

When we started writing this book, we did so in part for selfish reasons: as we have mentioned multiple times throughout the text, we teach an introductory course at our home institution, and we have never found a book that contains all of the information we feel is important for that course, at least not without a ton of other information and neither of us is a big fan of requiring a text of which we only intend to use a small part. Much like we discuss in Chapter 1, identifying needs and posing an appropriate problem statement are critical to the design process; similarly, identifying the areas we felt a text would augment our current introductory course helped shape the content that is included in this book. However, we did not want the book to read like a series of separate, independent topics or a compilation of short stories. As we have emphasized throughout Chapter 5, we wanted to ensure that the book maintained a level of continuity and that a story focusing on engineering success was woven throughout. Like any good story, this section is our effort to tie everything together and provide you with some means of reflecting upon how you might use your newfound skills moving forward.

Of course, this is not to say that these skills are not already being showcased in your classroom. Like a good book, your classes also tell a story, and although the individual topics or lessons may revolve around the development of specific skills, collectively, those skills are often complementary, and the grand finale of your time spent in that course might be a final project or assessment. These assignments are used by your instructors to demonstrate how the individual skills learned in the separate lessons can be put together, and they provide a means for the instructor to assess how well you have mastered them in an integrative fashion. In your class where this text is being used, we assume there is likely an overarching project, presentation, or some other means of demonstrating how you have been able to take the information provided within that course and integrate it in a meaningful way, similar to the synthesis of information discussed in Chapter 4.

Although we hope the skills discussed and developed throughout this book will serve you well throughout the rest of your academic and professional career, we would also like to think that many of the topics in this book are equally applicable to activities beyond the scope of academia and engineering. For example, the fundamental aspects of good communication described in Chapters 5 and 6 can easily extend beyond engineering and into your everyday life. Understanding your audience and tailoring your communication to those audiences fits with how you might talk, text, or share information with your parents, your grandparents, or friends. Similarly, a well-written letter or email can help resolve issues you might have with a service provider or merchant or might cheer up a friend or loved one. Granted, those may or may not contain equations and figures (we won't judge), but as we said back in Chapter 5, good writing is good writing regardless of the context.

Similarly, as we previously discussed in Chapter 4, information literacy is a skill that will serve you well for the rest of your life. Every citizen should be able to research a topic area, find information from valid sources, read those sources critically, and draw their own independent conclusion. As Mark Twain stated in his biography (although many attribute this quote to Twain, he directly attributes it to Benjamin Disraeli): "There are three kinds of lies: lies, damned lies, and statistics."[1] Perhaps that would have been a good quote to lead with in our section regarding statistics, but the purpose of that section wasn't to provide you with enough information to determine the type of statistical testing you might need to undertake to determine if your sample populations are statistically different (that is something we will leave to your statistics and additional upper-level math and engineering courses). Instead, it was meant to provide you with an appreciation as to why statistical analysis is important in testing and engineering design. Advertising is a prime example where facts can be easily manipulated, and without due diligence and appropriate research, many people blindly accept the information as fact. A classic example is Colgate toothpaste, which claimed in 2007 that "80% of dentists recommend" the toothpaste in their poster advertisement in the United Kingdom.[2] Many took that to read that four of five dentists surveyed selected Colgate as the top toothpaste. However, the reality was that a survey sent out to dentists asked them to recommend several toothpastes and brands, and Colgate appeared as one of the checked brands on 80% of the returned surveys. While stating that "80% of dentists recommend Colgate" was not completely false, the actual survey outcome was not as strong of an endorsement of the product as was implied in the advertisement.

Another area where information can be manipulated is when people attempt to demonstrate causation when, in reality, they are simply demonstrating correlation. If two events happen to occur at or near the same time, some people will use that information to demonstrate causation, even if it might be quite difficult (or even impossible) to demonstrate that one event caused the other. A classic example of this is the superstitions or rituals athletes will go through to recreate

---

1 Mark Twain, Chapters From My Autobiography, Project Gutenberg.
2 https://marketinglaw.osborneclarke.com/retailing/colgates-80-of-dentists-recommend-claim-under-fire/.

what they perceive to be working effectively when they or their team are on a winning streak. This might be lucky clothing (Michael Jordan wore the shorts from his time at North Carolina under his Bulls uniform during his entire professional career[3]), a pre-game ritual, or in-game activities. Even something as simple as a batter or pitcher undergoing the same ritual between pitches is an example of in-game activities that could be perceived to be helpful. The point of mentioning this is that even when the data appears to support someone's claim, it is best to do your own research and draw your own conclusions, provided the research is something you are capable of understanding. Perhaps if the debate is regarding quantum mechanics, you should defer to the experts[4] (this goes back to the engineering creeds we discussed in the Introduction). Suppose per capita consumption of mozzarella cheese correlates with the number of civil engineering doctorates awarded. That does not mean that civil engineering PhDs eat enough mozzarella cheese to shift the per capita eating habits in the United States, and any argument that suggests otherwise should be heavily scrutinized.[5]

One additional chapter that we hope you might find opportunities to integrate into other facets of your life is Chapter 2. While we wrote that chapter within the context of how to have success as an engineer, many of those ideas easily translate to life in general. Planning, time management, and SMART goals can be easily used for everything from your career to washing your car to doing chores around the house. Similarly, networking and finding activities you are personally interested in will allow you to meet and interact with a variety of people, increase your professional and social network, and expand your breadth and depth of knowledge, both inside and outside of engineering.

ENGINEERING SUCCESS STARTS WITH YOU

One activity we asked you to complete in Chapter 2 in regard to time management was to reflect on how your time was spent, including every activity you were involved in, both academic and personal. One thing the pandemic has taught many of us is the importance of self-appreciation and self-care. When we say engineering success starts with you, we mean that in order to be successful, you need to take care of yourself. This includes all aspects of your health and well-being: physical, mental, spiritual, intellectual, and social. While some of these may have been more difficult to pursue through the ebbs and flows of living through a global pandemic, recognizing that your well-being relies on each of these (perhaps some more than others depending on your own personality and obligations) should suggest that finding pursuits to fulfill your specific needs are critical for self-care. Recognize that it is often when one or more of these are out of balance

---

[3] https://abcnews.go.com/Technology/DyeHard/superstition-michael-jordans-success/story?id=11210274.
[4] https://www.nytimes.com/2019/09/07/opinion/sunday/quantum-physics.html.
[5] http://www.tylervigen.com/spurious-correlations.

that your ability to make ethical decisions is most likely to be compromised. Maintaining your health and well-being can help ensure that when you are asked to face a difficult decision, you can be confident that your response is one that you are much less likely to regret[6]. When looking at each of these aspects individually (although you will see that many overlap), realize that your university doubtless has resources available that you should take 100% advantage of during your time on campus.

- **Physical.** As we mention in Chapter 2, taking advantage of college intramurals, regular visitations to the campus recreation facilities, or simply connecting with your surroundings by regular walks or hikes are a great way to unplug, relax, and recharge. Additionally, paying attention to nutrition and sleep habits can help contribute to your overall physical well-being.
- **Mental.** Again, as we describe in Chapter 2, finding things you like to do outside of the daily grind of school can be beneficial to developing a well-rounded portfolio of interests but can also be a great way of recharging your focus and capacity to learn. Planned mental breaks, whether that be for exercise, hobbies, regular time with friends and/or family, or binge-watching your new favorite show on Netflix, are things people often do to decompress. Recognize, though, that sometimes these activities are not enough, and for many, the pandemic did not make anything easier. In fall 2020, the Jed Foundation found that 82% of students reported dealing with anxiety, 68% felt socially isolated or lonely, 63% reported being depressed, and 62% had trouble concentrating.[7] If one of these categories hits home, recognize first and foremost that you are certainly not alone in feeling this way, and furthermore, there is nothing "wrong" with you for feeling this way. Almost every campus has some type of student services to help students feeling this way, and we encourage you to find out more about your campus's mental health support services and get into contact with them if you are having difficulty coping with the stresses in your life. We will say to you what we say to our students - few individuals hesitate to take advantage of the library or campus recreation facilities; in the same way, you should not hesitate to take advantage of other support services offered on campus.
- **Spiritual.** Spiritual well-being can take a lot of different forms for different people. This could be being active in a campus group(s) that focuses on spirituality or becoming involved in a local/community group. For others, reconnecting with nature and natural surroundings will provide the same opportunity for comfort and/or renewal.
- **Intellectual.** We like to think about this from two different perspectives. One is where you are looking for opportunities or ways to expand your knowledge beyond what is provided to you in the engineering classroom, whether that is through taking additional classes that

---

6 https://hbr.org/2013/05/sleep-deprived-people-are-more-likely-to-cheat.

7 https://www.jedfoundation.org/survey-of-college-student-mental-health-in-2020/.

interest you, engaging in research activities, or simply learning new skills and facts on your own for your enjoyment and personal satisfaction rather than something that is required for your degree (although if you can find ways to blend these two items, all the better). The second perspective is for folks who perhaps are struggling with their classes for any number of reasons. Not to sound like a broken record, but as with our other bullet points, making sure you are taking advantage of all the resources available to you to ensure academic success is crucial to your future success. For example, as mentioned in Chapter 2, taking advantage of office hours might be one way to clarify any concepts that remain foreign to you after a lecture or lab. Alternatively, seeking out opportunities for help from teaching assistants, learning assistants, or other personnel in the class (laboratory personnel, staff, etc.) as well as working closely with your colleagues and peers can be another great resource. Beyond your own class, student organizations within your discipline might be a great way to introduce yourself to more senior students who have already completed the course you are in and may have been in your situation a few years earlier. Additionally, your university likely has other resources (see the section in Chapter 3) that may further help you in your academic pursuits or as you prepare for your career after graduation. Furthermore, centers for academic achievement, adaptive services, and other similarly named programs may provide you with the help you need, whether that is additional instruction, development of new learning skills to help you better learn and retain material, or help for overcoming learning disabilities that may be hindering you in the classroom.

- **Social.** Making sure you have found an outlet to spend time with friends and family may have been more difficult than usual recently. However, an era where "social distancing" has become part of our everyday vernacular, finding time to ensure you are maintaining healthy relationships with your family and friends should be one of your priorities. In addition, opportunities to engage with your peers can help at multiple levels with respect to this list. This is also where student organizations and other campus activities may play an important role in your well-being.

This list was based in part on our own experiences as undergraduate students, as long ago as that may have been (and no, we are not saying when) as well as the resources and organizations we have at our current home campus. As each university and college is different, we encourage you to explore what is available on your campus and how you can either get involved with or gain access to those entities. One thing we emphasize to our own students is that colleges and universities have a lot of activities and resources available to students (mentioned throughout this text), but it is up to the students to be proactive in participating in those activities or organizations or taking advantage of those resources. When you hear people talk about self-advocacy, this is exactly what they are trying to tell you: self-advocacy is the ability to communicate one's needs and wants and to make decisions about the support needed to achieve them.[8] Taking the time and effort to make sure you are in a good place to learn is foundational to ensuring your success in the classroom and beyond, and without such a foundation in place, the fruits of your efforts may not be fully actualized. In the end, be good to yourself and each other, and find the path in life that makes you happy.

REFERENCES

Twain, Mark. "Chapters From My Autobiography." Project Gutenberg. https://www.gutenberg.org/files/19987/19987-h/19987-h.htm.

Daly-Cano, Meada, Annemarie Vaccaro, and Barbara Newman. 2015. "College Student Narratives About Learning and Using Self-advocacy Skills." *Journal of Postsecondary Education and Disability* 28, no. 2: 213–27.

---

[8] Meada Daly-Cano, Annemarie Vaccaro, and Barbara Newman, "College Student Narratives About Learning and Using Self-advocacy Skills," Journal of Postsecondary Education and Disability.